Lecture Notes in Artificial Intelligence   10344

Subseries of Lecture Notes in Computer Science

LNAI Series Editors

Randy Goebel
   University of Alberta, Edmonton, Canada
Yuzuru Tanaka
   Hokkaido University, Sapporo, Japan
Wolfgang Wahlster
   DFKI and Saarland University, Saarbrücken, Germany

LNAI Founding Series Editor

Joerg Siekmann
   DFKI and Saarland University, Saarbrücken, Germany

More information about this series at http://www.springer.com/series/1244

Andreas Holzinger · Randy Goebel
Massimo Ferri · Vasile Palade (Eds.)

Towards Integrative Machine Learning and Knowledge Extraction

BIRS Workshop
Banff, AB, Canada, July 24–26, 2015
Revised Selected Papers

Springer

Editors
Andreas Holzinger (iD)
Medical University Graz
Graz
Austria

Randy Goebel (iD)
University of Alberta
Edmonton, AB
Canada

Massimo Ferri (iD)
Bologna University
Bologna
Italy

Vasile Palade (iD)
Coventry University
Coventry
UK

ISSN 0302-9743 ISSN 1611-3349 (electronic)
Lecture Notes in Artificial Intelligence
ISBN 978-3-319-69774-1 ISBN 978-3-319-69775-8 (eBook)
https://doi.org/10.1007/978-3-319-69775-8

Library of Congress Control Number: 2017957550

LNCS Sublibrary: SL7 – Artificial Intelligence

Cover illustration: Banff International Research Station for Mathematical Innovation and Discovery
Photograph taken by: Andreas Holzinger.

Printed on acid-free paper

This Springer imprint is published by Springer Nature
The registered company is Springer International Publishing AG
The registered company address is: Gewerbestrasse 11, 6330 Cham, Switzerland

Banff International Research Station
for Mathematical Innovation and Discovery

Preface

The BIRS Workshop 15w2181 in Banff was dedicated to stimulating a cross-domain integrative machine learning approach and appraisal of "hot topics" toward tackling the grand challenge of reaching a level of useful and useable computational intelligence with a focus on real-world problems, such as in the health domain. This encompasses learning from prior data, extracting and discovering knowledge, generalizing the results, fighting the curse of dimensionality, and ultimately disentangling the underlying explanatory factors in complex data, i.e., to make sense of data within the context of the application domain.

The workshop particularly tried to contribute advancements in promising novel areas as, e.g., at the intersection of machine learning and topological data analysis. History has shown that most often the overlapping areas at intersections of seemingly disparate fields are key for stimulation of new insights and further advances. This is particularly true for the extremely broad field of machine learning.

Successful machine learning needs a concerted effort, fostering integrative research between experts ranging from diverse disciplines – from data science to data visualization and always taking into account issues of privacy, data protection, safety, and security. Tackling such complex research undertakings needs both disciplinary excellence and cross-disciplinary networking without boundaries, and a cross-domain integration of experts – like what the international HCI-KDD group is doing now for many years. Consequently, we thank all our international colleagues who persistently energize our activities and support our general motto: "Science is to test crazy ideas – Engineering is to put these ideas into practice."

We are grateful for everybody who contributed directly or indirectly to this volume; in particular, we thank all our reviewers for their careful work and patience. Finally we want to say thank you to our families and friends for their personal support and last but not least we are grateful for the Springer management team and the Springer production team for their smooth and professional support!

September 2017

Andreas Holzinger
Randy Goebel
Vasile Palade
Massimo Ferri

Organization

Scientific Committee

Rakesh Agrawal	Microsoft Search Labs, Mountain View, USA
Beatrice Alex	University of Edinburgh, UK
Amin Anjomshoaa	Massachusetts Institute of Technology, USA
Joel P. Arrais	University of Coimbra, Portugal
John A. Atkinson-Abutridy	Universidad de Concepcion, Chile
Chloe-Agathe Azencott	Mines Paris Tech, France
Alexandra Balahur	European Commission Joint Research Centre, Ispra, Italy
Smaranda Belciug	University of Craiova, Romania
Mounir Ben Ayed	Ecole Nationale d'Ingenieurs de Sfax, Tunisia
Elisa Bertino	Purdue University, West Lafayette, USA
Chris Biemann	Technische Universität Darmstadt, Germany
Miroslaw Bober	University of Surrey, Guildford, UK
Francesco Buccafurri	Università Mediterranea di Reggio Calabria, Italy
Andre Calero-Valdez	RWTH Aachen University, Germany
Mirko Cesarini	Università di Milano Bicocca, Italy
Polo Chau	Georgia Tech, USA
Chaomei Chen	Drexel University, USA
Veronika Cheplygina	Erasmus Medical Center, The Netherlands
Nitesh V. Chawla	University of Notre Dame, USA
Alexiei Dingli	University of Malta, Malta
Sara Johansson Fernstad	Northumbria University, UK
Ana Fred	Technical University of Lisbon, Portugal
Aryya Gangopadhyay	UMBC Center of Cybersecurity, University of Maryland, USA
Panagiotis Germanakos	University of Cyprus, Cyprus
Michael Granitzer	University of Passau, Germany
Dimitrios Gunopulos	University of Athens, Greece
Helwig Hauser	University of Bergen, Norway
Andreas Hotho	University of Würzburg, Germany
Jun Luke Huan	University of Kansas, USA
Anthony Hunter	University College London, UK
Beatriz De La Iglesia	University of East Anglia, UK
Igor Jurisica	IBM Life Sciences Discovery Centre, and Princess Margaret Cancer Centre, Canada
Jiri Klema	Czech Technical University, Czech Republic

Peter Kieseberg	SBA Research gGmbH – Secure Business Austria, Austria
Negar Kiyavash	University of Illinois at Urbana-Champaign, USA
Lubos Klucar	Academy of Sciences, Slovakia
David Koslicki	Oregon State University, USA
Patti Kostkova	University College London, UK
Natsuhiko Kumasaka	RIKEN, Japan
Robert S. Laramee	Swansea University, UK
Nada Lavrac	Joszef Stefan Institute, Slovenia
Chunping Li	Tsinghua University, China
Haibin Ling	Temple University, USA
Luca Longo	Trinity College Dublin, Ireland
Lenka Lhotska	Czech Technical University Prague, Czech Republic
Andras Lukacs	Hungarian Academy of Sciences and Eoetvos University, Hungary
Avi Ma' Ayan	Mount Sinai Hospital, USA
Ljiljana Majnaric-Trtica	University of Osijek, Croatia
Vincenzo Manca	University of Verona, Italy
Ernestina Menasalvas	Polytechnic University of Madrid, Spain
Yoan Miche	Nokia Bell Labs, Finland
Antonio Moreno-Ribas	Intelligent Technologies for Advanced Knowledge
Katharina Morik	Technische Universität Dortmund, Germany
Abbe Mowshowitz	The City College of New York, USA
Marian Mrozek	Jagiellonian University, Poland
Zoran Obradovic	Temple University, USA
Daniel E. O'leary	University of Southern California, USA
Patricia Ordonez-Rozo	University of Puerto Rico Rio Piedras, Puerto Rico
Jan Paralic	Technical University of Kosice, Slovakia
Valerio Pascucci	University of Utah, USA
Gabriella Pasi	Università di Milano Bicocca, Italy
Armando J. Pinho	University of Aveiro, Portugal
Pavel Pilarczyk	Institute of Science and Technology Austria, Austria
Massimiliano Pontil	UCL London, UK
Raul Rabadan	Columbia University, USA
Heri Ramampiaro	Norwegian University of Science and Technology, Norway
Chandan K. Reddy	Wayne State University, USA
Gerhard Rigoll	Technische Universität München, Germany
Jianhua Ruan	University of Texas, USA
Lior Rokach	Ben-Gurion University of the Negev, Israel
Carsten Roecker	Fraunhofer IOSB-INA and Ostwestfalen-Lippe University of Applied Sciences, Germany
Timo Ropinski	Ulm University, Germany
Giuseppe Santucci	La Sapienza University of Rome, Italy
Pierangela Samarati	University of Milan, Italy
Michele Sebag	CNRS, Universite Paris Sud, France

Contents

About the Editors

Andreas Holzinger is lead of the Holzinger Group, HCI-KDD, Institute for Medical Informatics/Statistics at the Medical University Graz, and Associate Professor at the Institute of Interactive Systems and Data Science at Graz University of Technology. Currently, Andreas is Visiting Professor for Machine Learning in Health Informatics at the Faculty of Informatics at Vienna University of Technology. He serves as consultant for the Canadian, US, UK, Swiss, French, Italian, and Dutch Government, for the German Excellence Initiative, and as national expert in the European Commission. His research interests are in supporting human intelligence with machine intelligence to help to solve problems in health informatics. Andreas obtained a PhD in Cognitive Science from Graz University in 1998 and his Habilitation (second PhD) in Computer Science from Graz University of Technology in 2003. Andreas was Visiting Professor in Berlin, Innsbruck, London (twice), Aachen, and Verona. He founded the Expert Network HCI-KDD to foster a synergistic combination of methodologies of two areas that offer ideal conditions toward unraveling problems in understanding intelligence: Human–Computer Interaction (HCI) and Knowledge Discovery/Data Mining (KDD), with the goal of supporting human intelligence with machine learning. Andreas is Associate Editor of *Knowledge and Information Systems* (KAIS), Section Editor for machine learning of BMC *Medical Informatics and Decision Making* (MIDM), Editor of the journal *Machine Learning and Knowledge Extraction* (MAKE), organizer of the IFIP CD-MAKE conference, and member of IFIP WG 12.9 Computational Intelligence.

Randy Goebel is vice president of the Innovates Centre of Research Excellence (iCORE) and chair of the Alberta Innovates Academy, working as principal investigator at the Alberta Machine Intelligence Institute (amii) to enhance understanding and innovation in machine intelligence. Randy is vice president of research and professor of computer science at the Faculty of Sciences of the University of Alberta. He obtained a PhD in Computer Science from the University of British Columbia in 1985. Randy's theoretical work on abduction, hypothetical reasoning, and belief revision is internationally well known, and his recent application of practical belief revision and constraint programming to scheduling, layout, and Web mining is now having industrial impact. His recent research is focused on the application of machine learning to a variety of areas, including information extraction in health and medicine, and formalization of visualization. Application-specific induction projects include the creation of hypotheses on protein structure, and the automatic formation of semantic indexing structures used to support content-based retrieval of text and images. Randy has previously held faculty appointments at the University of Waterloo, Multimedia University (Malaysia), Hokkaido University (Japan), the University of Tokyo (Japan), and he is actively involved in academic and industrial collaborative research projects in Canada, Japan, China, and at the German Research Center for Artificial Intelligence in Saarbrücken.

Vasile Palade is a Reader in the Faculty of Engineering and Computing at Coventry University, UK. Vasile is a well-acknowledged international expert in machine learning. He previously held academic and research positions at the University of Oxford - UK, University of Hull - UK, and the University of Galati - Romania. Vasile's research interests lie in the area of machine learning/computational intelligence, and encompass neural and neuro-fuzzy systems, various nature-inspired algorithms such as swarm optimization algorithms, hybrid intelligent systems, among others. He holds a PhD in intelligent systems from the University of Galati, Romania. Vasile's application areas include bioinformatics and medical informatics problems, fault diagnosis, and web usage mining, among others. Vasile is an Associate Editor for several international journals, such as Knowledge and Information Systems, International Journal on Artificial Intelligence Tools, International Journal of Hybrid Intelligent Systems, Neurocomputing, and others. Vasile is an IEEE Senior Member and a member of the IEEE Computational Intelligence Society.

Massimo Ferri is head of the Vision Mathematics group and Full Professor of Geometry at the Department of Mathematics at the Engineering Faculty of Bologna University (the oldest university in the world, in continuous operation since 1088 A.D.). Massimo started his research in the topology of manifolds; in more recent years he drifted toward applications of geometry and topology to some aspects of robotics, in particular computer vision. He worked on the project ADAM of the European Union for early automatic diagnosis of melanoma and the project "Geometry and Topology of Robot Vision" of the Italian Space Agency to mention only a few. Recently, he is interested in the intersection of geometry and machine learning. Massimo was Visiting Professor at a number of international institutions including the University of Warwick, UK, the University of Erlangen-Nuremberg, Germany, the University Federal de Pernambuco, Recife, Brazil, and the University of Florence, Italy. Massimo was a member of the Scientific Committee of the Italian Mathematical Union for the periods 2000–2003 and 2006–2009 and he is a reviewer for several mathematics and computer visions journal.

Towards Integrative Machine Learning and Knowledge Extraction

Andreas Holzinger[1]([✉]), Randy Goebel[2], Vasile Palade[3], and Massimo Ferri[4]

[1] Holzinger Group, HCI-KDD, Institute for Medical Informatics/Statistics,
Medical University Graz, Graz, Austria
`a.holzinger@hci-kdd.org`
[2] Centre for Machine Learning, University of Alberta, Edmonton, Canada
`rgoebel@ualberta.ca`
[3] Cogent Computing Applied Research Centre, Coventry University, Coventry, UK
`vasile.palade@coventry.ac.uk`
[4] Vision Mathematics Group, Department of Mathematics,
University of Bologna, Bologna, Italy
`massimo.ferri@unibo.it`

Abstract. This Volume is a result of workshop 15w2181 "Advances in interactive knowledge discovery and data mining in complex and big data sets" at the Banff International Research Station for Mathematical Innovation and Discovery. The workshop was dedicated to bring together experts with diverse backgrounds but with one common goal: *to understand intelligence* for the successful design, development and evaluation of *algorithms* that can learn from data, extract knowledge from experience, and to improve their learning behaviour over time – similarly as we humans do. Knowledge discovery, data mining, machine learning, artificial intelligence are more or less synonymously used with no strict definitions or boundaries. "Integrative" means to support not only the machine learning & knowledge extraction pipeline, ranging from dealing with data in arbitrarily high-dimensional spaces to the visualization of results into a lower dimension accessible to a human; it is taking into account seemingly disparate fields which can be very fruitful when brought together - for solving problems in complex application domains (e.g. health informatics). Here we want to emphasize that the most important findings in machine learning will be those we do not know yet. In this paper we provide: (1) a short motivation for the integrative approach; (2) brief summaries of the presentations given in Banff; and (3) some personally flavoured, subjective future research outlooks, e.g. in the combination of geometrical approaches with machine learning.

Keywords: Integrative machine learning · Knowledge discovery

1 Introduction and Motivation

Machine learning deals with *understanding intelligence* for the design, development and evaluation of algorithms that can learn from data to gain knowledge

© Springer International Publishing AG 2017
A. Holzinger et al. (Eds.): Integrative Machine Learning, LNAI 10344, pp. 1–12, 2017.
https://doi.org/10.1007/978-3-319-69775-8_1

from experience and to improve their learning behaviour over time, similarly as we humans do [1,2]. The challenge is to discover *relevant* structural and/or temporal patterns ("knowledge") in data, thus machine learning is inherently connected to knowledge extraction [3].

Why do we use the extension "integrative"? Machine learning has many theoretical aspects and is deeply grounded in the field of artificial intelligence (AI) [4,5], however, we want to emphasize that machine learning is a very practical field with many diverse application areas (e.g. health informatics). We are of the opinion that problem solving in such domains need an integrative/integrated approach. The meaning of the words integrative or integrated stems from Latin *integratus*, which means "make whole", i.e. *"to put together parts or elements and combine them into a harmonious, interrelated whole, so that constituent units function in a cooperatively manner"*.

Integrative Machine Learning is based on the idea of combining the best of the two worlds dealing with understanding intelligence, which is manifested in the HCI–KDD approach: [6–8]: Human–Computer Interaction (HCI), rooted in cognitive science, particularly dealing with *human intelligence*, and Knowledge Discovery/Data Mining (KDD), rooted in computer science particularly dealing with *computational intelligence* [9]. This approach fosters a complete machine learning and knowledge extraction (MAKE) pipeline, ranging from the very physical issues of data pre-processing, mapping and fusion of arbitrarily high-dimensional data sets up to the visualization of the results in a dimension accessible to a human end-user and making data interactively accessible and manipulable. Among the greatest application challenges is health informatics, which is not surprising, because health is a good example for a domain full of uncertainty and complex problems, where we are constantly confronted with probabilistic, unknown, incomplete, heterogenous, noisy, dirty, erroneous, inaccurate, and missing data in arbitrarily high dimensional spaces [10,11]. Moreover, the health domain requires issues of privacy, data protection, safety and security, along with trust, acceptance and social issues - which are also included in the integrative machine learning approach [12].

2   Presentations

ANDREAS HOLZINGER: *Challenges of biomedicine, health and the life sciences and the chances of Interactive Machine Learning for Knowledge Discovery.* Andreas opened the workshop with providing an overview of the variations and enormous complexity and heterogeneity of data sets from the health domain and the challenges researchers are faced [10]. He emphasized that automatic machine learning algorithms aiming at bringing the human-out-of-the-loop [13] have demonstrated impressive success in various domains. Particularly, deep learning, supported by cloud-CPUs and large data sets can exceed human performance in visual tasks, playing games [14,15] and even in medical classification tasks [16]. Andreas showed examples of Gaussian processes, where automatic approaches (e.g., kernel machines) struggle on function extrapolation problems, which are astonishingly trivial for humans. Consequently, interactive machine learning, by integrating a *human-in-the-loop*, thereby making use of human cognitive abilities, seems to be a promising approach for the near future. This is particularly useful to solve problems with complex data and/or rare events, where traditional learning algorithms (e.g., deep learning) suffer of insufficient training samples and a human-in-the-loop can help to solve problems which otherwise would remain NP-hard [17].

RANDY GOEBEL: *The role of logic and machine learning within a general theory of visualization.* Randy pointed out that the role of logic and machine learning [3] in visualization is not familiar to many colleagues, but the idea of visual inference requires inductive transformations from base data to visual data. Randy emphasized that these transformations need to be constrained by inference principles, including the construction of layers of knowledge, which generally are difficult to construct by hand. The idea is to describe how logic, learning, and visualization are connected, in order to help enable humans to make better inferences from growing volumes of data in every area of application. Moreover, he pointed out that visualization is an abstraction process, and that abstractions from partial information, however voluminous they might be, directly confronts the non monotonic reasoning challenge; thus the need for caution in engineering visualization systems without carefully considering the consequences of visual abstraction. This is particularly important with interactive visualization, which has recently formed the basis for such fields as visual analytics and are a natural bridge to machine learning [18].

VASILE PALADE: *Class Imbalance Learning.* Vasile demonstrated that class imbalance of data is commonly found in many data mining tasks and machine learning applications to real-world problems. When learning from imbalanced data, the performance measure used for model selection plays a vital role. The existing and popular performance measures used in class imbalance learning, such as the Gm and Fm, can still result in sub-optimal classification models. Vasile first presented a new performance measure, called the Adjusted Geometric-mean (AGm), which overcomes the problems of the existing performance measures when learning from imbalanced data. Support Vector Machines (SVMs) has become a very popular and effective machine learning technique, but which can still produce sub-optimal models when it comes to imbalanced data sets. Vasile presented then FSVM-CIL (Fuzzy SVM for Class Imbalance Learning), an effective method to train FSVMs with imbalanced data in the presence of outliers and noise in the data. Finally, Vasile discussed some efficient re-sampling methods for training SVMs with imbalance data in the context of applications [19].

KATHARINA MORIK: *Big Data and Small Devices.* Katharina showed that big data are produced by various sources. Most often, they are distributedly stored at computing farms or clouds. Analytics on the Hadoop Distributed File System (HDFS) then follows the MapReduce programming model (batch layer). It is complemented by the speed layer, which aggregates and integrates incoming data streams in real time. Katharina emphasized that when considering big data and small devices, obviously, we imagine the small devices being hosts of the speed layer, only. Analytics on the small devices is restricted by memory and computation resources. The interplay of streaming and batch analytics offers a multitude of configurations. The collaborative research center SFB 876 investigates data analytics for and on small devices regarding runtime, memory and energy consumption. Katharina investigated in her talk graphical models, which generate the probabilities for connected (sensor) nodes. Resource-restricted methods deliver insights fast enough for a more interactive analysis [20].

SIBYLLE HESS: *Investigation of Code Tables to compress and describe the underlying characteristics of binary databases.* Sibylle is a young Computer Science student and inspected the spectrum of methods (from frequent pattern mining to numerical optimization) to extract the pattern set that describes a binary database best. Invoking the Minimum Description Length (MDL) principle, this objective can be stated as: find the code table that compresses the database most. Sibylle pointed out that a particularly interesting interpretation of this task, relating it to biclustering, arises from the formulation as a matrix factorisation problem. Finally, Sibylle stressed that biclustering has a variety of applications in research fields such as collaborative filtering, gene expression analysis and text mining. She emphasized that the derived matrix factorisation analogy provides a new perspective on distinct data mining subfields (unifying biclustering and pattern mining concepts such as Krimp), initialising a cross-over of their applications and interpretations of derived models [21].

KATHARINA HOLZINGER: *Darwin, Lamarck, Baldwin, Mendel: What can we learn from them?.* Katharina is a young student of Natural Sciences and discussed the potential of evolutionary algorithms, inspired by biological mechanisms observed in nature, such as selection and genetic changes, to find the best solution for a given optimisation problem [22]. Contrary to Darwin, and according to Lamarck and Baldwin, organisms in natural systems learn to adapt over their lifetime and allow to adjust over generations. Whereas earlier research was rather reserved, more recent research underpinned by the work of Lamarck and Baldwin, finds that these theories have much potential, particularly in upcoming fields relevant for health informatics, such as epigenetics. Katharina emphasized particularly the integration of the Theories of Gregor Mendel, which could be helpful for extending machine learning techniques [23].

NITESH CHAWLA: *Big Data and Small Data for Personalized and Population Health care.* Nitesh showed that proactive personalized medicine can bring fundamental changes in health care. He asked the question: "Can we then take a data-driven approach to discover nuggets of knowledge and insight from the big data in health care for patient-centered outcomes and personalized health care?" and Nitesh asked if we may answer the question: "What are my disease risks and how to best manage it?". Particularly he pointed out the importance of the question: "How to scale this at the population level?". Nitesh discussed some work of his group that takes the data and networks driven thinking to personalized health care and patient-centered outcomes [24]. He demonstrated the effectiveness of population health data to drive personalized disease management and wellness strategies, and in effect impacting population health. Nitesh also shared various pilots under-way that take the algorithms and tools on a "road-show". In this volume Nitesh and his group contribute on the impact of complex health care data on the Machine Learning pipeline.

YUZURU TANAKA: *Exploratory Visual Analytics for the Discovery of Complex Analysis Scenarios for Big Data.* Yuzuru emphasized that the data-centric approach is increasing its significance in varieties of scientific research areas and large-scale social cyber-physical systems. He showed examples from disparate areas: Biomedical research and urban-scale winter road management. Through his involvement in three major projects on these subjects, Yuzuru recognized a big gap between the state-of-the-art big data core technologies and both the data-centric research for the analysis of clinico-genomic trial data and the big data approach to the optimization of social system services. During the last decades, enabling core technologies for big data analysis have made remarkable advances in both analysis and management technologies, however, he pointed out that we still lack methodologies to find out the best analysis scenario for finding out such solutions as personalized medicines from a given clinico-genomic trial data set. He also proposed exploratory visual analytics to support analysts to find out complex analysis scenarios, and the coordinated multiple views and analyses framework as its application framework [25].

MATEUSZ JUDA: *Homology of big data - algorithms and applications.* Mateusz started with demonstrating homology as a well known and powerful tool in pure mathematics and he stressed that for many years it was impossible to use this tool in applied science because of data size and cubical algorithms for computing homology. He explained that new preprocessing methods give us now a possibility to apply homology for real data, e.g. from sensor networks [26]. Discrete Morse theory is an example of such a tool, which simplifies data without changing its topological information. Mateusz introduced discrete Morse theory and its application to homology computations and he showed how to construct a discrete vector field (Morse matching) using parallel and distributed algorithms. Mateusz also showed an application of this tool to knots detection and classification in a biological context.

MASSIMO FERRI: *Persistent topology for natural shape analysis and image retrieval.* Massimo emphasized that data are often of "natural" origin (pictures or 3D meshes representing living beings, faces, handwritten words, hand-drawn sketches etc.), and that classic mathematical techniques do not fit well the task of analyzing, comparing, classifying, retrieving such data. On the contrary topology (and in particular algebraic topology) is, by its very nature, the part of mathematics which formalizes qualitative aspects of objects; therefore topological data processing and topological data mining well integrate with more classical mathematical tools. Massimo then concentrated on persistent homology, which combines geometry and algebraic topology in the study of pairs (X, f) where X

is an object (typically a topological space) and f is a continuous function defined on X (typically with real values). One application is the extraction of topological features of an object out of a cloud of sample points. Another class of applications uses f as a formalization of a classification criterion; in this case various functions can give different criteria, cooperating in a complex classifier. Massimo explained that persistent homology is studied by several teams throughout the world and has already given rise to several applications: dermatological diagnosis, evolution of hurricanes, signature recognition, gesture recognition; retrieval of trademarks, 3D meshes, hand-drawn sketches etc. To this volume Massimo contributes with a survey on persistent topology for natural data analysis.

MIRKO CESARINI: *Data Quality in Schema free (big) data.* Mirko focused in his presentation on the challenges and open problems emerging when complex data sets are used to obtain insights about a population e.g., analysing job offers using data from web job boards, inspecting the job history of the working population (starting from administrative records), and analysing cellular network traffic. He pointed out that a huge set of weakly structured data can be derived from information sources containing a variety of data types. Mirko explained that in such a context, techniques ranging from formal methods to machine learning can identify and exploit information structures (both hidden and visible) to check data consistency, to ameliorate the data (e.g., fixing inconsistencies), and to create synthetic representations of the original data.

SOU-CHENG CHOI: *Machine Learning for Machine Data in Computational Social Sciences.* In this last talk of the workshop, Sou-Cheng presented machine learning and high-accuracy prediction methods of rare events in semi-structured or unstructured log files produced at high velocity and high volume by NORC's computer-assisted telephone interviewing network. These machine log files are generated by their internal Voxco Servers for a telephone survey. Sou-Cheng and her colleagues adapt natural language processing (NLP) techniques and data-mining methods to train powerful learning and prediction models for error messages in the absence of source code, updated documentation, and relevant dictionaries. Such approaches can be useful for applications in other domains, e.g. the biomedical domain. In this volume Sou-Cheng contribute with a comparison of public-domain software and services for probabilistic record linkage and address standardization. Additionally, to these selected and peer reviewed papers, this volume contains some more selected and peer reviewed papers: Andreas Holzinger and his group provide an overview to machine learning for digital pathology; Fahrnaz Jayrannejad and Tim Conrad report on better interpretable models for proteomics data analysis using rule-based mining; Arnaud Nguembang Fadja and Fabrizio Riguzzi discuss probabilistic logic programming; Jefferson Tales Oliva and Joao Luis Garcia Rosa demonstrate predictive models for differentiation between normal and abnormal EEG through cross-correlation and machine learning techniques; Vincenzo Manca gives a brief philosophical note on information; The Rabadan group contributes with a paper on a fast semi-automatic segmentation tool for processing brain tumor images; Taimanov and his group are contributing with a paper on topological characteristics of oil and gas reservoirs and their applications, and finally Singh et al. are reporting on current experiments with deep learning approaches for the application in ambient assisted living.

3 Future Outlook

There are many future research directions in machine learning generally, and in the combination of machine learning with other approaches, e.g., in bridging probabilistic approaches with classic ontological approaches, or with topological approaches [12]. Here we present briefly some incomplete, subjective and personally flavoured research directions which we found interesting during our joint discussions in Banff. It should be noted that our applications domains are mostly health and health related areas, but not exclusively.

3.1 Persistence

Similarity is a concept which sounds very natural to a human being, but is very difficult to formalize for use in a machine; it is anyway of paramount importance in data retrieval and data mining. The most widely accepted formalization is by defining a group of transformations such that two objects, which can be mapped into each other by a transformation of the group, are considered to be similar. The classical geometrical transformation groups generally suffer from a rigidity, which can only be smoothened by a heavy use of statistics. Topological transformations (*homeomorphisms*), on the other hand, are too "free". Another problem is that different observers might have different similarity concepts depending, e.g., on their specific tasks.

Applications suggested to adapt topology — and in particular its branch called *homology* — to take the observer's viewpoint into account and to restrict consequently the set of transformations.

The main idea was to convey the observer's viewpoint into a function, called *measuring* (now *filtering*) function (in what was then called "size theory" [27,28]). The object is no more only a set — or, more precisely, a topological space — but a pair (X, f) of a space X and a function f defined on X and with real values. The filtration given by the sublevel sets of f then moderates the "freedom" of the topological setting; moreover, the possible use of different functions on the same space gives the method a powerful modularity. The theory was extended by what is now called *persistent homology* [29,30]. A thorough survey on (1-dimensional) persistence is [31]. In the last few years the extension of the theory to filtering functions with multidimensional range is the object of a hard investigation [32–35].

Keystones of persistent homology are:

– *Persistence diagrams* For each nonnegative integer i, there is a persistence diagram consisting of a set of points in the plane, which condenses the essential information on the pair (X, f) (through the homology modules of degree i of the sublevel sets of the filtration given by f).

- *Natural pseudodistance* Given pairs (X, f) and (Y, g), there is a way to measure how much a homeomorphism from X to Y distorts the filtrations given by f and g. The minimum of such measures among all possible homeomorphism is a (pseudo)distance which formalizes the dissimilarity of the two pairs by a number.
- *Matching distance* Given the persistence diagrams (of the same degree i) of two pairs (X, f) and (Y, g), there is a well-defined distance between the diagrams which is an optimal lower bound for the natural pseudodistance of the pairs.

Challenges in Persistence. Multidimensional ranges for filtering functions may be crucial for applications. There are examples showing that shape comparison by a multidimensional filtering function is finer than the use of the separate components [36,37]. There are anyway theoretical and computational hurdles that the community is trying to overcome.

The matching distance is too heavy to compute in data mining. Preprocessing by a sloppier but much faster distance is necessary. This seems to be the case if persistence diagrams are transformed into complex polynomials and the distance is computed on their coefficients; preliminary results support this strategy [38].

Most important, the choice of invariance groups and an interactive selection of filtering functions (or components thereof) are a special benefit of persistence; they promise to be of great advantage for relevance feedback in a human-in-the-loop scheme [39].

3.2   Evolutionary Algorithms for Big Data Processing

Evolutionary computing algorithms can be used for solving various optimization tasks as part of the solution for complex problems that involve big data and high dimensionality. Another recent employment of evolutionary algorithms [40], with very good potential application to big data processing, is to solve data sampling problems. Sampling is a basic and one of the most important tasks in data processing, statistics and machine learning, and acquiring good samples is not an easy task for an arbitrary probability distribution of data or when the data space is huge. Evolutionary Sampling proposed in [41] combines the popular rejection sampling method with other strategies within a probabilistic framework in order to obtain an optimal approximation of any pointwise computable density function by using finite samples, which is a fundamental problem in statistics and machine learning area. The paper also argues that many machine learning problems can be described as, or could be converted into corresponding, density function approximation problem problems, where the Evolutionary Sampling approach can be employed for training and as a machine learning method not as a mean for acquiring data only. Theoretical and experimental studies have demonstrated that the Evolutionary Sampling learning can be used to solve many practical application problems which can be expressed as density function approximation problems within a probabilistic framework.

Challenges and Future Work

- Investigating the role of evolutionary computing techniques when dealing with optimization and learning problems involving big data, such as for dynamic and very high dimensional problems, multi-objective big data analytics problems, or big data driven optimization of complex systems, is a very good and promising avenue for future research, which has not been properly tackled yet in the literature.
- Developing new nature inspired evolutionary algorithms. Investigating the performance (convergence and time complexity) of our previous swarm intelligent algorithms, such as the Random Drift Particle Swarm Optimization [42], and Quantum Particle Swarm Optimization variants [43], and apply them to big data and complex optimisation problems from bioinformatics and computational biology.
- Investigate the performance of the Evolutionary Sampling method in solving big data problems, especially from the biological and medical domain.

3.3 Graphical Models for Big Data

Probabilistic and graphical models are popular tools to use when processing large scale data sets today, and Bayesian networks is undoubtedly the most common approach within this family. However, inferring large Bayesian networks from data, such as inferring Genetic Regulatory Networks from genomic time series data, is a very challenging task in machine learning today. The current algorithms for inferring a Bayesian network from data, irrespective of the application area involved, work for small networks, of less than 100 nodes or so. But, the smallest networks we find in biology and medicine are at least an order of magnitude bigger; for example, the genetic regulatory network for the smallest organisms in nature and which are commonly studied by researchers in biology today has at least several thousands of nodes.

Challenges and Future Work

- Developing efficient algorithms for inferring large Bayesian networks from data, by extending the work in [44].
- Investigating the performance of the Dense Structural Expectation Maximisation algorithm proposed in [44] for large network inference problems, especially on networks generated from biology area, such as genetic regulatory networks, but not only.

Acknowledgements. We are grateful to all participants of the Banff BIRS workshop 15w2181, specifically to our colleagues from the international HCI-KDD expert network and generally to all colleagues who constantly support our group in fostering the idea of an integrated machine learning approach and in bringing together diverse areas in an cross-disciplinary manner to stimulate fresh ideas and to encourage multidisciplinary problem solving. The past has shown that many new discoveries are made in overlapping areas of seemingly disjunct fields and the interesting and most important discoveries are those which we have not yet found.

References

1. Bengio, Y.: Learning deep architectures for AI. Found. Trends Mach. Learn. **2**, 1–127 (2009)
2. Gopnik, A., Glymour, C., Sobel, D.M., Schulz, L.E., Kushnir, T., Danks, D.: A theory of causal learning in children: causal maps and bayes nets. Psychol. Rev. **111**, 3–32 (2004)
3. Poole, D., Mackworth, A., Goebel, R.: Computational Intelligence: A Logical Approach. Oxford University Press, New York (1998)
4. Michalski, R.S., Carbonell, J.G., Mitchell, T.M.: Machine Learning: An Artificial Intelligence Approach. Springer, Heidelberg (1983). doi:10.1007/978-3-662-12405-5
5. Ghahramani, Z.: Probabilistic machine learning and artificial intelligence. Nature **521**, 452–459 (2015)
6. Holzinger, A.: On knowledge discovery and interactive intelligent visualization of biomedical data - challenges in human computer interaction & biomedical informatics. In: Helfert, M., Fancalanci, C., Filipe, J. (eds.) DATA 2012, International Conference on Data Technologies and Applications, pp. 5–16 (2012)
7. Holzinger, A.: Human-computer interaction and knowledge discovery (HCI-KDD): what is the benefit of bringing those two fields to work together? In: Cuzzocrea, A., Kittl, C., Simos, D.E., Weippl, E., Xu, L. (eds.) CD-ARES 2013. LNCS, vol. 8127, pp. 319–328. Springer, Heidelberg (2013). doi:10.1007/978-3-642-40511-2_22
8. Holzinger, A.: Trends in interactive knowledge discovery for personalized medicine: cognitive science meets machine learning. IEEE Intell. Inform. Bull. **15**, 6–14 (2014)
9. Holzinger, A., Jurisica, I.: Knowledge discovery and data mining in biomedical informatics: the future is in integrative, interactive machine learning solutions. In: Holzinger, A., Jurisica, I. (eds.) Interactive Knowledge Discovery and Data Mining in Biomedical Informatics. LNCS, vol. 8401, pp. 1–18. Springer, Heidelberg (2014). doi:10.1007/978-3-662-43968-5_1
10. Holzinger, A., Dehmer, M., Jurisica, I.: Knowledge discovery and interactive data mining in bioinformatics - state-of-the-art, future challenges and research directions. BMC Bioinf. **15**, I1 (2014)
11. Lee, S., Holzinger, A.: Knowledge discovery from complex high dimensional data. In: Michaelis, S., Piatkowski, N., Stolpe, M. (eds.) Solving Large Scale Learning Tasks. Challenges and Algorithms. LNAI, vol. 9580, pp. 148–167. Springer, Cham (2016). doi:10.1007/978-3-319-41706-6_7
12. Holzinger, A.: Introduction to machine learning and knowledge extraction (make). Mach. Learn. Knowl. Extr. **1**, 1–20 (2017)
13. Shahriari, B., Swersky, K., Wang, Z., Adams, R.P., de Freitas, N.: Taking the human out of the loop: a review of bayesian optimization. Proc. IEEE **104**, 148–175 (2016)
14. Mnih, V., Kavukcuoglu, K., Silver, D., Rusu, A.A., Veness, J., Bellemare, M.G., Graves, A., Riedmiller, M., Fidjeland, A.K., Ostrovski, G., Petersen, S., Beattie, C., Sadik, A., Antonoglou, I., King, H., Kumaran, D., Wierstra, D., Legg, S., Hassabis, D.: Human-level control through deep reinforcement learning. Nature **518**, 529–533 (2015)
15. Silver, D., Huang, A., Maddison, C.J., Guez, A., Sifre, L., van den Driessche, G., Schrittwieser, J., Antonoglou, I., Panneershelvam, V., Lanctot, M., Dieleman, S., Grewe, D., Nham, J., Kalchbrenner, N., Sutskever, I., Lillicrap, T., Leach, M., Kavukcuoglu, K., Graepel, T., Hassabis, D.: Mastering the game of go with deep neural networks and tree search. Nature **529**, 484–489 (2016)

16. Esteva, A., Kuprel, B., Novoa, R.A., Ko, J., Swetter, S.M., Blau, H.M., Thrun, S.: Dermatologist-level classification of skin cancer with deep neural networks. Nature **542**, 115–118 (2017)
17. Holzinger, A., Plass, M., Holzinger, K., Crisan, G.C., Pintea, C.M., Palade, V.: A glass-box interactive machine learning approach for solving np-hard problems with the human-in-the-loop. arXiv:1708.01104 (2017)
18. Goebel, R.: Why visualization is an ai-complete problem (and why that matters). In: 20th International Conference on Information Visualisation (IV 2016), pp. 27–32. IEEE (2016)
19. Lopez, V., Fernandez, A., García, S., Palade, V., Herrera, F.: An insight into classification with imbalanced data: empirical results and current trends on using data intrinsic characteristics. Inf. Sci. **250**, 113–141 (2013)
20. Piatkowski, N., Lee, S., Morik, K.: Integer undirected graphical models for resource-constrained systems. Neurocomputing **173**, 9–23 (2016)
21. Hess, S., Morik, K., Piatkowski, N.: The primping routine-tiling through proximal alternating linearized minimization. Data Min. Knowl. Disc. **31**, 1090–1131 (2017)
22. Holzinger, K., Palade, V., Rabadan, R., Holzinger, A.: Darwin or lamarck? Future challenges in evolutionary algorithms for knowledge discovery and data mining. In: Holzinger, A., Jurisica, I. (eds.) Interactive Knowledge Discovery and Data Mining in Biomedical Informatics: State-of-the-Art and Future Challenges. LNCS, vol. 8401, pp. 35–56. Springer, Heidelberg (2014)
23. Holzinger, A., Blanchard, D., Bloice, M., Holzinger, K., Palade, V., Rabadan, R.: Darwin, lamarck, or baldwin: applying evolutionary algorithms to machine learning techniques. In: Slezak, D., Dunin-Keplicz, B., Lewis, M., Terano, T. (eds.) IEEE/WIC/ACM International Joint Conferences on Web Intelligence (WI) and Intelligent Agent Technologies (IAT), pp. 449–453. IEEE (2014)
24. Nagrecha, S., Thomas, P.B., Feldman, K., Chawla, N.V.: Predicting chronic heart failure using diagnoses graphs. In: Holzinger, A., Kieseberg, P., Tjoa, A.M., Weippl, E. (eds.) CD-MAKE 2017. LNCS, vol. 10410, pp. 295–312. Springer, Cham (2017). doi:10.1007/978-3-319-66808-6_20
25. Sjöbergh, J., Kuwahara, M., Tanaka, Y.: Visualizing clinical trial data using pluggable components. In: 2012 16th International Conference on Information Visualisation (IV), pp. 291–296. IEEE (2012)
26. Dlotko, P., Ghrist, R., Juda, M., Mrozek, M.: Distributed computation of coverage in sensor networks by homological methods. Appl. Algebra Eng. Commun. Comput. **23**(1/2), 1–30 (2012). doi:10.1007/s00200-012-0167-7
27. Frosini, P.: Measuring shapes by size functions. In: Intelligent Robots and Computer Vision X: Algorithms and Techniques, International Society for Optics and Photonics, pp. 122–133 (1992)
28. Verri, A., Uras, C., Frosini, P., Ferri, M.: On the use of size functions for shape analysis. Biol. Cybern. **70**, 99–107 (1993)
29. Edelsbrunner, H., Letscher, D., Zomorodian, A.: Topological persistence and simplification, pp. 454–463 cited By 72 (2000)
30. Carlsson, G., Zomorodian, A., Collins, A., Guibas, L.J.: Persistence barcodes for shapes. Int. J. Shape Model. **11**, 149–187 (2005)
31. Edelsbrunner, H., Harer, J.: Persistent homology-a survey. Contemp. Math. **453**, 257–282 (2008)
32. Frosini, P., Mulazzani, M.: Size homotopy groups for computation of natural size distances. Bull. Belg. Math. Soc. Simon Stevin **6**, 455–464 (1999)
33. Carlsson, G., Zomorodian, A.: The theory of multidimensional persistence. Discrete Comput. Geom. **42**, 71–93 (2009)

34. Biasotti, S., Cerri, A., Frosini, P., Giorgi, D., Landi, C.: Multidimensional size functions for shape comparison. J. Math. Imaging Vis. **32**, 161–179 (2008)
35. Cerri, A., Di Fabio, B., Ferri, M., Frosini, P., Landi, C.: Betti numbers in multidimensional persistent homology are stable functions. Math. Methods Appl. Sci. **36**, 1543–1557 (2013)
36. Cagliari, F., Di Fabio, B., Ferri, M.: One-dimensional reduction of multidimensional persistent homology. Proc. Am. Math. Soc. **138**, 3003–3017 (2010)
37. Adcock, A., Rubin, D., Carlsson, G.: Classification of hepatic lesions using the matching metric. Comput. Vis. Image Underst. **121**, 36–42 (2014)
38. Di Fabio, B., Ferri, M.: Comparing persistence diagrams through complex vectors (2015)
39. Frosini, P.: G-invariant persistent homology. Math. Methods Appl. Sci. **38**, 1190–1199 (2015)
40. Xie, Z., Sun, J., Palade, V., Wang, S., Liu, Y.: Evolutionary sampling: a novel way of machine learning within a probabilistic framework. Inf. Sci. **299**, 262–282 (2015)
41. Jun, S., Palade, V., Xiao-Jun, W., Wei, F., Zhenyu, W.: Solving the power economic dispatch problem with generator constraints by random drift particle swarm optimization. IEEE Trans. Ind. Inform. **10**, 222–232 (2014)
42. Jun, S., Palade, V., Xiaojun, W., Wei, F.: Multiple sequence alignment with hiddenmarkov models learned by random driftparticle swarm optimization. IEEE/ACM Trans. Comput. Biol. Bioinform. **11**, 243–257 (2014)
43. Sun, J., Fang, W., Palade, V., Wu, X., Xu, W.: Quantum-behaved particle swarm optimization with gaussian distributed local attractor point. Appl. Math. Comput. **218**, 3763–3775 (2011)
44. Fogelberg, C., Palade, V.: Dense structural expectation maximisation with parallelisation for efficient large-network structural inference. Int. J. Artif. Intell. Tools **22**, 1350011 (2013)

Machine Learning and Knowledge Extraction in Digital Pathology Needs an Integrative Approach

Andreas Holzinger[1]([✉]), Bernd Malle[1,2], Peter Kieseberg[1,2], Peter M. Roth[3], Heimo Müller[1,4], Robert Reihs[1,4], and Kurt Zatloukal[4]

[1] Holzinger Group, HCI-KDD, Institute for Medical Informatics/Statistics, Medical University Graz, Graz, Austria
andreas.holzinger@medunigraz.at
[2] SBA Research, Vienna, Austria
[3] Institute of Computer Graphics and Vision, Graz University of Technology, Graz, Austria
pmroth@icg.tugraz.at
[4] Institute of Pathology, Medical University Graz, Graz, Austria
kurt.zatloukal@medunigraz.at

Abstract. During the last decade pathology has benefited from the rapid progress of image digitizing technologies, which led to the development of scanners, capable to produce so-called Whole Slide images (WSI) which can be explored by a pathologist on a computer screen comparable to the conventional microscope and can be used for diagnostics, research, archiving and also education and training. Digital pathology is not just the transformation of the classical microscopic analysis of histological slides by pathologists to just a digital visualization. It is a disruptive innovation that will dramatically change medical work-flows in the coming years and help to foster personalized medicine. Really powerful gets a pathologist if she/he is augmented by machine learning, e.g. by support vector machines, random forests and deep learning. The ultimate benefit of digital pathology is to enable to learn, to extract knowledge and to make predictions *from a combination of heterogenous data,* i.e. the histological image, the patient history and the *omics data. These challenges call for integrated/integrative machine learning approach fostering transparency, trust, acceptance and the ability to explain step-by-step *why a decision has been made.*

Keywords: Digital pathology · Data integration · Integrative machine learning · Deep learning · Transfer learning

1 Introduction and Motivation

The ability to mine "sub-visual" image features from digital pathology slide images, features that may not be visually discernible by a pathologist, offers the opportunity for better quantitative modeling of disease appearance and

© Springer International Publishing AG 2017
A. Holzinger et al. (Eds.): Integrative Machine Learning, LNAI 10344, pp. 13–50, 2017.
https://doi.org/10.1007/978-3-319-69775-8_2

hence possibly improved prediction of disease aggressiveness and patient outcome. However, the compelling opportunities in precision medicine offered by big digital pathology data come with their own set of computational challenges. Image analysis and computer assisted detection and diagnosis tools previously developed in the context of radiographic images are woefully inadequate to deal with the data density in high resolution digitized whole slide images. Additionally, there has been recent substantial interest in combining and fusing radiologic imaging, along with proteomics and genomics based measurements with features extracted from digital pathology images for better prognostic prediction of disease aggressiveness and patient outcome. Again there is a paucity of powerful tools for combining disease specific features that manifest across multiple different length scales. The purpose of this paper is to discuss developments in computational image analysis tools for predictive modeling of digital pathology images from a detection, segmentation, feature extraction, and tissue classification perspective. We discuss the emergence of new handcrafted feature approaches for improved predictive modeling of tissue appearance and also review the emergence of deep learning schemes for both object detection and tissue classification. We also briefly review some of the state of the art in fusion of radiology and pathology images and also combining digital pathology derived image measurements with molecular "omics" features for better predictive modeling [1].

The adoption of data-intensive methods can be found throughout various branches of health, leading e.g. to more evidence-based decision-making and to help to go towards personalized medicine [2]: A grand goal of future biomedicine is to tailor decisions, practices and therapies to the individual patient. Whilst personalized medicine is the ultimate goal, stratified medicine has been the current approach, which aims to select the best therapy for groups of patients who share common biological characteristics. Here, ML approaches are indispensable, for example *causal inference trees (CIT)* and aggregated grouping, seeking strategies for deploying such stratified approaches. Deeper insight of personalized treatment can be gained by studying the personal treatment effects with *ensemble CITs* [3]. Here the increasing amount of heterogenous data sets, in particular "-omics" data, for example from genomics, proteomics, metabolomics, etc. [4] make traditional data analysis problematic and optimization of knowledge discovery tools imperative [5,6]. On the other hand, many large data sets are indeed large collections of small data sets. This is particularly the case in personalized medicine where there might be a large amount of data, but there is still a relatively small amount of data for each patient available [7]. Consequently, in order to customize predictions for each individual it is necessary to build a model for each patient along with the inherent uncertainties, and to couple these models together in a hierarchy so that information can be "borrowed" from other similar patients. This is called *model personalization*, and is naturally implemented by using hierarchical Bayesian approaches including e.g. hierarchical Dirichlet processes [8] or Bayesian multi-task learning [9].

This variety of problems in Digital Pathology requires a synergistic combination of various methodological approaches which calls for a combination of various approaches, e.g. geometrical approaches with deep learning models [10].

After a short motivation and explanation of why machine aided pathology is interesting, relevant and important for the future of diagnostic medicine, this article is organized as follows:

In Sect. 2 we provide a glossary of the most important terms.

In Sect. 3 we give an overview of where digital pathology is already in use today, which technologies of slide scanning are currently state-of-the-art, and describe the next steps towards a machine aided pathology. A sample use-case shall demonstrate the typical work-flow. Because data-integration, data fusion and data-preprocessing is an important aspect, we briefly describe these issues here.

In Sect. 4 we describe the most promising state-of-the-art machine learning technologies which can be of use for digital pathology.

In Sect. 5, finally, we discuss some important future challenges in machine learning, which includes multi-task learning, transfer learning and the use of multi-agent-hybrid systems.

2   Glossary and Key Terms

Automatic Machine Learning (aML) in bringing the human-out-of-the-loop is the grand goal of ML and works well in many cases having "big data" [11].

Big Data is indicating the flood of data today; however, large data sets are necessary for aML approaches to learn effectively; the problem is in "dirty data" [12], and sometimes we have large collections of little, but complex data.

Data Fusion is the process of integration multiple data representing the same real-world object into one consistent, accurate, and useful representation.

Data Integration is combining data from different sources and providing a unified view.

Deep Learning allows models consisting of multiple layers to learn representations of data with multiple levels of abstraction [13].

Digital Pathology is not only the conversion of histopathological slides into a digital image (WSI) that can be uploaded to a computer for storage and viewing, but a complete new medical work procedure.

Dimensionality of data is high, when the number of features p is larger than the number of observations n by magnitudes. A good example for high dimensional data is gene expression study data [14].

Explainability is motivated due to lacking transparency of black-box approaches, which do not foster trust and acceptance of ML among end-users. Rising legal and privacy aspects, e.g. with the new European General Data Protection Regulations, make black-box approaches difficult to use, because they often are not able to explain why a decision has been made [15].

interactive Machine Learning (iML) in bringing the human-in-the-loop is beneficial when having small amounts of data ("little data"), rare events or dealing with complex problems [16,17], or need reenactment (see explainability).

Knowledge Discovery (KDD) includes exploratory analysis and modeling of data and the organized process to identify valid, novel, useful and understandable patterns from these data sets [18].

Machine Aided Pathology is the management, discovery and extraction of knowledge from a virtual case, driven by advances of digital pathology supported by feature detection and classification algorithms.

Multi-Task Learning (MTL) aims to learn a problem together with multiple, different but related other problems through shared parameters or a shared representation. The underlying principle is *bias learning* based on probable approximately correct learning (PAC learning) [19].

Topological Data Mining uses algebraic geometry to recover parameters of mixtures of high-dimensional Gaussian distributions [20].

Transfer Learning How can machine learning algorithms perform a task by exploiting knowledge, extracted during solving previous tasks? Contributions to solve this problem would have major impact to Artificial Intelligence generally, and Machine Learning specifically [21].

Virtual Case is the set of all histopathological slides of a case together with meta data from the macro pathological diagnosis [22].

Virtual Patient has very different definitions (see [23], we define it as a model of electronic records (images, reports, *omics) for studying e.g. diseases.

Visualization can be defined as transforming the symbolic into the geometric and the graphical presentation of information, with the goal of providing the viewer with a qualitative understanding of the information contents [6,24].

Whole Slide Imaging (WSI) includes scanning of all tissue covered areas of a histopathological slide in a series of magnification levels and optional as a set of focus layers.

3   From Digital Pathology to Machine Aided Pathology

3.1   Digital Pathology

Modern pathology was founded by RUDOLF VIRCHOW (1821-1902) in the mid of the 19th century. In his collection of lectures on Cellular Pathology (1858) he set the basis of modern medical science and established the "microscopically thinking" still applied today by every pathologist. In histopathology a biopsy or surgical specimen is examined by a pathologist, after the specimen has been processed and histological sections have been placed onto glass slides. In cytopathology either free cells (fluids) or tissue micro-fragments are "smeared" on a slide without cutting a tissue.

In the end of the 20th century an individual clinical pathologist was no longer able to cover the knowledge of the whole scientific field. This led to today's specialization of clinical pathology either by organ systems or methodologies. Molecular biology and *omics technologies set the foundation for the emerging field of molecular pathology, which today alongside WSI provides the most important source of information, especially in the diagnosis of cancer and infectious diseases.

The roots of digital pathology go back to the 1960s, when first telepathology experiments took place. Later in the 1990s the principle of virtual microscopy [25] appeared in several life science research areas. At the turn of the century the scientific community more and more agreed on the term "digital pathology" [26] to denote digitization efforts in pathology.

However in 2000 the technical requirements (scanner, storage, network) were still a limiting factor for a broad dissemination of digital pathology concepts. Over the last 5 years this changed as new powerful and affordable scanner technology as well as mass/cloud storage technologies appeared on the market. This is also clearly reflected in the growing number of publications mentioning the term "digital pathology" in PMC, see Fig. 1.

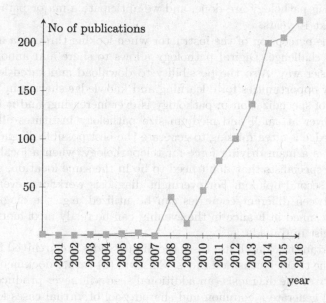

Fig. 1. Number of publication in PMC containing the term "digital pathology".

The field of Radiology has undergone the digital transformation almost 15 years ago, not because radiology is more advanced, but there are fundamental differences between digital images in radiology and digital pathology: The image source in radiology is the (alive) patient, and today in most cases the image is even primarily captured in digital format. In pathology the scanning is done from preserved and processed specimens, for retrospective studies even from slides stored in a biobank. Besides this difference in pre-analytics and metadata content, the required storage in digital pathology is two to three orders of magnitude higher than in radiology. However, the advantages anticipated through digital pathology are similar to those in radiology:

Capability to transmit digital slides over distances quickly, which enables telepathology scenarios.

Capability to access past specimen from the same patients and/or similar cases for comparison and review, with much less effort then retrieving slides from the archive shelves.

Capability to compare different areas of multiple slides simultaneously (slide by slide mode) with the help of a virtual microscope.

Capability to annotate areas directly in the slide and share this for teaching and research.

Digital pathology is today widely used for educational purposes [27] in telepathology and teleconsultation as well as in research projects. Digital pathology in diagnostics is an emerging and upcoming field. With the validation of the first WSI systems for primary diagnosis by the FDA the first steps for the digital transition in pathology are done, and we anticipate a major paradigm shift within the next 10 years.

Sharing the perception of the instructor when looking through a microscope is a technical challenge. Digital pathology allows to share and annotate slides in a much easier way. Also the possibility to download annotated lecture sets generates new opportunities for e-learning and knowledge sharing in pathology.

The level of specialization in pathology is ever increasing, and it is no more possible to cover at small and medium size pathology institutes all fields, so expert knowledge is often missing to generate the best possible diagnosis for the patient. This is a main driving force for telepathology, when a local team can easily involve specialists that don't need to be in the same location, and/or get a specialized second opinion. For overnight diagnosis workflows even the time difference between different countries can be utilized, e.g. the diagnosis for a virtual case scanned in France in the evening can be ready next morning, done by a pathologist in Canada.

It is important that in all use cases the digital slides are archived in addition to the analogue tissue blocks and slides. This will (a) ensure documentation and reproducibility of the diagnosis (an additional scan will never produce the same WSI) and (b) generate a common and shared pool of virtual cases for training and evaluation of machine learning algorithms. Archiving WSI is even a prerequisite for the validation and documentation of diagnostic workflows, especially when algorithmic quantification and classification algorithms are applied. In the next sections we describe requirements for data management and digital slide archiving as a starting point for machine aided pathology scenarios.

3.2 Virtual Case

A pathological workflow always starts with the gross evaluation of the primary sample. Depending on the medical question and the material type small tissue parts are extracted from the primary sample and are either embedded in a paraffin block or cryo-frozen. From the tissue blocks the pathology labs cuts several slides, applies different staining methods and conducts additional histological and molecular tests. Finally, the pathologists evaluate all the slides together with the supporting gross-and molecular findings and makes the diagnosis. If in

ICD-10 Diagnosis		Slides	FSA	MOD
H60.4	Cholesteatoma of external ear	1	no	no
K37	Unspecified appendicitis	2	no	no
K21	Gastro-esophageal reflux disease w. esophagitis	4	no	no
K52.9	Noninfective gastroenteritis and colitis	6	no	no
C67.9	Malignant neoplasm of bladder	8	yes	no
C34	Neoplasm of bronchus or lung	10	yes	yes
I51.7	Cardiomegaly	12	no	no
C56	Malignant neoplasm of ovary	14	yes	no
N60.3	Fibrosclerosis of breast	16	yes	no
C85.9	Malignant lymphoma, non-Hodgkin, NOS	18	yes	yes
C18	Malignant neoplasm of colon	20	yes	yes
C92.1	Chronic myeloid leukemia	22	no	no
N40	Benign prostatic hyperplasia	25	no	no
C61	Malignant neoplasm of prostate	36	yes	no
D07.5	Carcinoma in situ of prostate	43	no	no
C83.5	Lymphoblastic (diffuse) lymphoma	50	yes	no

Fig. 2. Average number of slides for different pathological diagnosis. FSA: frozen section analysis; MOD: molecular diagnosis. Source: Analysis of all findings in the year 2016 at the Institute of Pathology, Graz Medical University.

addition to the set of WSI all information is present in a structured digital format, we call this a virtual case. In a virtual case, the average number of slides and additional findings varies very much for different medical questions and material types. Figure 2 shows the average number of slides for different diagnosis done in the year 2016 at the Institute of Pathology at Graz Medical University.

$$15 \times 15\,\text{mm@}0.12\mu m/pixel = 125000 \times 125000 = 15.6 Gigapixel \quad (1)$$
$$15.6 Gigapixel@3 \times 8bit/pixel = 46.9\,\text{GB}(uncompressed) \quad (2)$$
$$46.9\,\text{GB} \div 3(jpeg2000) = 15.6\,\text{GB}(lossless) \quad (3)$$
$$46.9\,\text{GB} \div 20(jpeg2000, highQ) = 2.3\,\text{GB}(lossy) \quad (4)$$
$$46.9\,\text{GB} \div 64(jpeg2000, mediumQ) = 0,7\,\text{GB}(lossy) \quad (5)$$

The most demanding data elements in a virtual case are the whole slide images (WSI). Compared to radiology, where the typical file size are between 131 KB for MRI images, 524 KB for CT-Scans, 18 MB for digital radiology, 27 MB for digital mammography and 30 MB for computed radiography [28], a single WSI scan with 80x magnification consists of 15.6 Gigapixels. For the calculation of the WSI file size and comparison of different scanner manufacturers, we use the de-facto standard area of 15 mm x 15 mm, with an optical resolution of 0.12 μm, which corresponds to an 80x magnification (see Fig. 3).

With 8 bit information for each color channel a WSI results in 46.9 GB stored in an uncompressed image format. Looking at the number of slides of a typical case, it is clear, that some compression techniques must be applied to the image data, and luckily several studies reported that lossy compression with a high

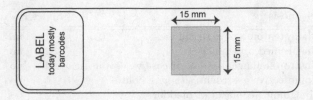

Fig. 3. Schematic view of a histopathological slide. An area of 15 mm x 15 mm is the de-facto standard for quoting scan speed and size.

quality level does not influence the diagnostic results. Still there are unresolved questions:

High compression levels. Can the compression level be increased up to 200 without significant influence in human decision making, e.g. with optimized *jpeg2000* algorithms and intra-frame compression techniques for z-layers.

Tissue/Staining dependencies. Does the maximum compression level depend on tissue type and staining?

Compression in ML scenarios. What compression level can be applied when WSI images are part of machine learning training sets and/or classified by algorithms?

The newest generation of scanners (as of 2017 !) is able to digitize a slide at various vertical focal planes, called z-layers, each the size of a singe layer. The multi-layer image can be either combined by algorithms to a single composite multi-focus image (Z-stacking) or used to simulate the fine focus control of a conventional microscope. Z-stacking is a desirable feature especially when viewing cytology slides, however, the pathologist should be aware that such an image can never be seen through the microscope (see Fig. 4).

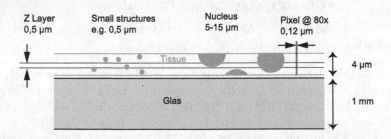

Fig. 4. Focus layers in a typical histopathological slide, thickness 4 μm.

At the Institute of Pathology at Graz Medical University, which is a medium to large organization, about 73,000 diagnosis are made within a year and approx 335,000 glass slides are produced in the pathology lab, approx 25,000 glass slides in the cytology lab. This results in a yearly storage capacity of almost 1 PetaByte and the appropriate computing power to process approx. 1000 slides per day plus the necessary capacity to train and improve ML algorithms. This numbers illustrate that the digital transformation of diagnostic workflows in pathology will demand for very high storage, even when stored in a compressed format, as well as computing capacity.

Several data formats are used today, either vendor independent (DICOM, TIFF/BigTIFF, Deep Zoom images) and vendor specific formats from Aperio, Hamamatsu, Leica, 3DHistech, Philips, Sakura and Trestle. In the setup of a virtual slide archive for medical research and machine learning it is essential to (a) agree on a common exchange format, and (b) to separate patient related and image related metadata. Patient related metadata comprise direct identifiers (name, birthday, zip code, ...) but also diagnosis results and others results from the patient medical history. When no such data is stored within or attached to the image format, the WSI is purely anonymous, as no re-identification of the patient is possible. To link between the same WSI used in different studies, either a global unique identifier (GUID) or a image generated hash can be used.

3.3   Towards Machine Aided Pathology

Digitizing workflows is one important enabling step to a groundbreaking change in clinical pathology, where AI methods and ML paradigms are introduced to pathological diagnosing. This assistance starts with simple classification and quantification algorithms as already available today, and ends in a full autonomous pathologist, where human expertise is replaced by machine intelligence. To distinguish such scenarios from simple digital workflows we propose the term **machine aided pathology**, when a significant contribution of the decision making process is supported by machine intelligence. Machine aided pathology solutions can be applied at several steps of the diagnosis making process:

Formulation of a hypothesis. Each diagnosis starts with a medical question and a corresponding underlying initial hypothesis. The pathologist refines this hypothesis in an iterative process, consequently looking for known patterns in a systematic way in order to confirm, extend or reject his/her initial hypothesis. Unconsciously, the pathologist asks the question *"What is relevant?"* and zooms purposefully into the -according to his/her opinion - essential areas of the cuts. The duration and the error rate in this step vary greatly between inexperienced and experienced pathologists. An algorithmic support in this first step would contribute in particular to the quality and interoperability of pathological diagnoses and reduce errors at this stage, and would be particularly helpful for educational purposes. A useful approach is known from Reeder and Felson (1975) [29] to recognize so called gamuts in images and to interpret these according to the most likely and most unlikely, an approach having its origin in differential diagnosis.
  - Very large amounts of data can only be managed with a "multi resolution" image processing approach using image pyramids. For example, a Colon cancer case consists of approximately 20 Tera (!) pixel of data - a size which no human is capable of processing.

– The result of this complex process is a central hypothesis, which has to be tested on a selection of relevant areas in the WSI, which is determined by quantifiable values (receptor status, growth rate, etc.).

– Training data sets for ML can now contain human learning strategies (transfer learning) as well as quantitative results (hypotheses, areas, questions, etc.).

Detection and classification of known features. Through a precise classification and quantification of selected areas in the sections, the central hypothesis is either clearly confirmed or rejected. In this case, the pathologist has to consider that the entire information of the sections is no longer taken into account, but only areas relevant to the decision are involved. It is also quite possible that one goes back to the initial hypothesis step by step and changes their strategy or consults another expert, if no statement can be made on the basis of the classifications.

– In this step ML algorithms consist of well known standard classification and quantification approaches.

– An open question is how to automatically or at least semi-automatically produce training sets, because here specific annotations are needed (which could come from a stochastic ontology, e.g.).

– Another very interesting and important research question is, whether and to what extent solutions learned from one tissue type (organ 1) can be transferred to another tissue type (organ 2) – transfer learning – and how robust the algorithms are with respect to various pre-analytic methods, e.g. stainings, etc.

Risk prediction and identification of unknown features. Within the third step, recognized features (learned parameters) are combined to a diagnosis and an overall prediction of survival risk. The main challenge in this step lies in training/validation and in the identification of novel, previously unknown features from step two. We hypothesize that the pathologist supported by machine learning approaches is able to discover patterns – which previously were not accessible! This would lead to new insights into previously unseen or unrecognized relationships.

Besides challenges in ML, also the following general topics and prerequisites have to be solved for a successful introduction of machine aided pathology:

Standardization of WSI image formats and harmonization of annotation/metadata formats. This is essential for telepathology applications and even more important for the generation of training sets, as for a specific organ and disease stages, even at a large institute of pathology the required amount of cases may not be available.

Common digital cockpit and visualization techniques should be used in education, training and across different institutes. Changing the workplace should be as easy as switching the microscope model or manufacturer. However, commonly agreed-upon visualization and interaction paradigms can only be achieved in a cross vendor approach and with the involvement of the major professional associations.

3.4 Data Integration

The image data (see Fig. 5) can be fused with two other sources of data: (1) Clinical data from electronic patient records [30], which contain documentations, reports, but also laboratory tests, physiological parameters, recorded signals, ECG, EEG, etc.); this also enables linking to other image data (standard X-ray, MR, CT, PET, SPECT, microscopy, confocal laser scans, ultrasound imaging, molecular imaging, etc.) (2) *omics data [4], e.g. from genomic sequencing technologies (Next Generation Sequencing, NGS, etc.), microarrays, transcriptomic technologies, proteomic and metabolomic technologies, etc., which all plays important roles for biomarker discovery and drug design [31,32].

Data integration is a hot topic in health informatics generally and solutions can bridge the gap between clinical routine and biomedical research [33]. This is becoming even more important due to the heterogeneous and different data sources, including picture archiving and communication systems (PACS) and radiological information systems (RIS), hospital information systems (HIS), laboratory information systems (LIS), physiological and clinical data repositories, and all sorts of -omics data from laboratories, using samples from biobanks. Technically, *data integration* is the combination of data from different sources

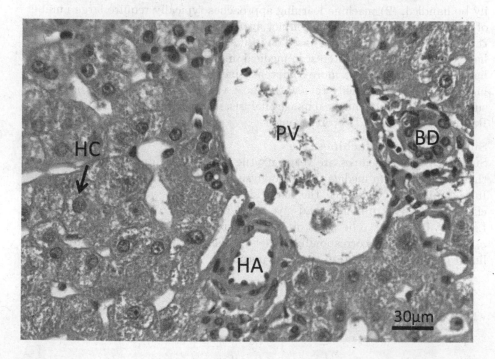

Fig. 5. Detail of a typical WSI: Hematoxylin and eosinstained histological section of a formalin-fixed and paraffin-embedded normal human liver tissue. Manual annotation: PV, portal vein; BD, bilde duct; HA, hepatic artery, HC (arrow), example of hepatocyte. Bar = 30 μm (Image Source: Pathology Graz)

and providing users with a unified view on these data, whereas *data fusion* is matching various data sets representing one and the same object into a single, consistent and clean representation [34]; in health informatics these unified views are particularly important in high-dimensions, e.g. for integrating heterogeneous descriptions of the same set of genes [35]. The general rule is that fused data is more informative than the original separate inputs. Inclusion of these different data sources and a fresh look on the combined views would open future research avenues [36].

4   Machine Learning in Medical Image Analysis

Computer-added diagnosis has become an important tool in medicine to support medical doctors in their daily life. The general goals are to classify images to automatically detect diseases or to predict the healing process. Thus, medical imaging builds on several low level tasks such as segmentation, registration, tracking and detection. Many of these tasks can be efficiently solved via machine learning approaches, where, in contrast to typical computer vision problem, we are facing several problems: (1) medical image data such as obtained from CT, MR, or X-ray show specific characteristics (e.g., blur and noise) that cannot easily be handled; (2) machine learning approaches typically require large number of training samples, which is often not available; (3) there are no clear labels as the ground truth is often just based on visual inspection by humans. Thus, there has been a considerable interest in medical image analysis and many approaches have been proposed. As a more comprehensive discussion would be out-of-scope, in the following, we briefly review the most versatile and tools that have been successfully applied in medical image analysis, namely, Support Vector Machines, Random Forests, and Deep Learning.

Support Vector Machines
Support Vector Machines are very versatile tools in machine learning and have thus also be used in medical image analysis for different tasks and applications. In the following, we sketch the main ideas, where we will focus on the two-class classification problem, and give a brief summary of related applications. Let $\mathcal{L} = \{(\mathbf{x}_i, y_i)\}_{i=1}^{L}$ be a set of pairs, where $\mathbf{x}_i \in \mathbb{R}^N$ are input vectors and $y_i \in \{+1, -1\}$ their corresponding labels. Then the objective is to determine a linear classification function (i.e., a hyperplane)

$$f(\mathbf{x}) = \mathbf{w}^{\top}\mathbf{x}_i + b, \tag{6}$$

where $\mathbf{w} \in \mathbb{R}^N$, and b is a bias term, such that

$$\mathbf{w}_i^{\top}\mathbf{x} + b \begin{cases} > 0 & \text{if } y_i = 1 \\ < 0 & \text{if } y_i = -1, \end{cases} \tag{7}$$

which is equivalent to

$$y_i(\mathbf{w}^{\top}\mathbf{x}_i + b) > 0, \ i = 1, \dots, L. \tag{8}$$

If the training data is linear separable, then there will exist an infinite number of hyperplanes satisfying Eq. (8). To ensure a unique solution and to increase the linear separability for unseen data (i.e., reduce the generalization error), support vector machines build on the concept of the margin (which is illustrated in Fig. 6), which is defined as the minimal perpendicular distance between a hyperplane and the closest data points. In particular, the decision boundary is chosen such that the margin M is maximized. By taking into account the relation $\|\mathbf{w}\| = \frac{1}{M}$, the maximum margin can be obtained my minimizing $\|\mathbf{w}\|^2$:

$$\min_{\mathbf{w},b} \frac{1}{2}\|\mathbf{w}\|^2$$
$$\text{s.t. } y_i(\mathbf{w}^\top \mathbf{x}_i + b) \geq 1, \ i = 1, \ldots, L. \tag{9}$$

In order to solve the constrained problem Eq. (9) for $\mathbf{w}$ and b, we introduce the Lagrange multipliers $\beta_i, i = 1, \ldots, L$, and use the Kuhn-Tucker theorem to convert the problem to the unconstrained dual problem (Wolfe dual):

$$\max \sum_{i=1}^{L} \beta_i - \frac{1}{2} \sum_{i}^{L} \sum_{j}^{L} \beta_i \beta_j y_i y_j \mathbf{x}_i^\top \mathbf{x}_j$$
$$\text{s.t. } \sum_{i=1}^{L} \beta_i y_i = 0, \quad \beta_i \geq 0 \quad i = 1, \ldots, L. \tag{10}$$

In this way, we get the decision function $\hat{f}$ for classifying unseen observations $\mathbf{x}$ as

$$\hat{f}(\mathbf{x}) = \text{sign}\left(\mathbf{w}^\top \mathbf{x} + b\right), \tag{11}$$

which is equivalent to

$$\hat{f}(\mathbf{x}) = \text{sign}\left(\sum_{i}^{L} \beta_i y_i \mathbf{x}^\top \mathbf{x}_i + b\right), \tag{12}$$

where $\beta_i > 0$ if $\mathbf{x}_i$ is on the boundary of the margin, and $\beta_i = 0$ otherwise. Thus, it can be seen that $\mathbf{w}$ can be estimated only via a linear combination of samples on the boundary, which are referred to as support vectors (see also Fig. 6).

If the data is not linearly separable, we can apply the kernel trick. As can be seen, Eqs. (10) and (12), the data does only appear in form of dot products $\langle \mathbf{x}_i, \mathbf{x}_i \rangle = \mathbf{x}_i^\top \mathbf{x}_j$. When introducing a transformation

$$\Phi(\cdot): \mathbb{R}^N \to \mathbb{R}^P, \tag{13}$$

we need only to estimate the dot product $\langle \Phi(\mathbf{x}_i), \Phi(\mathbf{x}_j) \rangle = \Phi(\mathbf{x}_i)^\top \Phi(\mathbf{x}_j)$. Thus, if there is a kernel function

$$K(\mathbf{x}_i, \mathbf{x}_j) = \langle \Phi(\mathbf{x}_i), \Phi(\mathbf{x}_j) \rangle \tag{14}$$

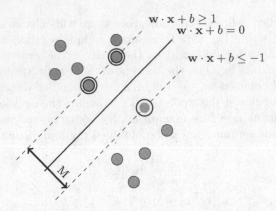

$$w \cdot x + b \geq 1$$
$$w \cdot x + b = 0$$
$$w \cdot x + b \leq -1$$

Fig. 6. Maximal margin for an SVM: The decision boundary for the two classes (red and blue balls) is estimated such that the margin M is maximized. The samples on the margin-boundary (indicated by the black ring) are referred to as support vectors. (Color figure online)

the dot product can be estimated without explicitly knowing Φ. Moreover, any other valid kernel can be used, for example:

- Linear kernel: $K(\mathbf{x}_i, \mathbf{x}_j) = \mathbf{x}_i^\top \mathbf{x}_j$,
- Polynomial kernel: $K(\mathbf{x}_i, \mathbf{x}_j) = \left(\mathbf{x}_i^\top \mathbf{x}_j + 1\right)^d$,
- Radial Basis Function (RBF) Kernel: $K(\mathbf{x}_i, \mathbf{x}_j) = e^{-\gamma \|\mathbf{x}_i - \mathbf{x}_j\|^2}$,
- Mahalanobis kernel: $K(\mathbf{x}_i, \mathbf{x}_j) = e^{-(\mathbf{x}_i - \mathbf{x}_j)^\top \mathbf{A}(\mathbf{x}_i - \mathbf{x}_j)}$.

In this way Eqs. (10) and (12) can be generalized to

$$L_D(\boldsymbol{\beta}) = \sum_{i=1}^m \beta_i - \frac{1}{2} \sum_i^m \sum_j^m \beta_i \beta_j y_i y_j K(\mathbf{x}_i, \mathbf{x}_j) \tag{15}$$

and

$$\hat{f}(\mathbf{x}) = \text{sign}\left(\sum_i^m \beta_i y_i K(\mathbf{x}_i, \mathbf{x}_j) + b \right). \tag{16}$$

Besides the flexibility to chose an appropriate kernel for a specific application, it is straightforward to extend the standard formulation for overlapping class distribution by introducing the concept of soft margins. In addition, there exist several ways to extend the standard formulation to multiple classes (e.g., one-vs.-all SVM, pairwise SVM, and error-correcting-output code SVM), to apply SVMs for regression tasks, or to use it in the context of online/incremental and semi-supervised learning. In this way, SVMs are very flexible and widely applicable for the highly diverse task to be solved in medical imaging. For a more detailed review, we like to refer to [37–39]).

One of the most important application in medical imaging is to segment and classify image regions. For example, in [40] SVMs are used to segment lesions in ultrasound images. In particular, a kernel SVM using an RBF-kernel is used to segment both ultrasound B-mode and clinical ultrasonic images. Similarly, in [41] an effective retinal vessel segmentation technique is presented, allowing to drastically reduce the manual effort of ophthalmologists. To this end, first features are extracted which are then classified using a linear SVM. This makes not only the evaluation very fast, but also allows to learn a model form a smaller training set.

A different approach is followed in [42] to segment blood vessels based on fully connected conditional random fields. However, an efficient inference approach is applied, which is learned via a Structured Output SVM. In this way, a fully automated system is obtained that achieves human-like results. Similarly, [43] presents a fully automatic method for brain tissue segmentation, where the goal is to segment 3D MRI images of brain tumor patients into healthy and tumor areas, including their individual sub-regions. To this end, an SVM classification using multi-spectral intensities and textures is combined with a CRF regularization.

A slightly different application in medical imaging is to localize image regions. For example, [44] presents an approach to detect microcalcification (MC) clusters in digital mammograms via an SVM-based approach. This is in particular of interest, as MC clusters can be an early indicator for female breast cancer. A different application, but a similar approach was discussed in [45]. The goal is to localize the precise location of cell nuclei, helping in an automated microscopy applications such as such as cell counting and tissue architecture analysis. For this purpose three different inputs are used (i.e., raw pixel values, edge values, and the combination of both), which are used to train an SVM classifier based on an RBF-kernel.

Random Forests

Random Forests (RFs) [46], in general, are ensembles of decision trees, which are independently trained using randomly drawing samples from the original training data. In this way, they are fast, easy to parallelize, and robust to noisy training data. In addition, they are very flexible, paving the way for classification, regression, and clustering tasks, thus making them a valid choice for a wide range of medical image applications [47].

More formally, Random Forests are ensembles of T decision trees $\mathcal{T}_t(\mathbf{x})$: $\mathcal{X} \rightarrow \mathcal{Y}$, where where $\mathcal{X} = \mathbb{R}^N$ is the N-dimensional feature space and $\mathcal{Y}$ is the label space $\mathcal{Y} = \{1, \ldots, C\}$. A decision tree can be considered a directed acyclic graph with two different kinds of nodes: internal (split) nodes and terminal (leaf) nodes. Provided a sample $\mathbf{x} \in \mathcal{X}$, starting from the root node at each split node a decision is made to which child node the sample should be send, until it reaches a leave node. Each leaf note is associated with a model that assigns an input $\mathbf{x}$ an output $y \in \mathcal{Y}$. Each decision tree thus returns a class probability $p_t(y|\mathbf{x})$ for a given test sample $\mathbf{x} \in \mathbb{R}^N$, which is illustrated in Fig. 7(b). These probabilities are then averaged to form the final class probabilities of the RF. A class decision for a sample $\mathbf{x}$ is finally estimated by

$$y^* = \arg\max_y \frac{1}{T} \sum_{t=1}^{T} p_t(y|\mathbf{x}). \tag{17}$$

During training of a RF, each decision tree is provided with a random subset of the training data $\mathcal{D} = \{(x_1, y_1), \dots (x_{|D|}, y_{|D|})\} \subseteq \mathcal{X} \times \mathcal{Y}$ (i.e., bagging) and is trained independently from each other. The data set $\mathcal{D}$ is then recursively split in each node, such that the training data in the newly created child nodes is pure according to the class labels. Each tree is grown until some stopping criterion (e.g., a maximum tree depth) is met and class probability distributions are estimated in the leaf nodes. This is illustrated in Fig. 7(a).

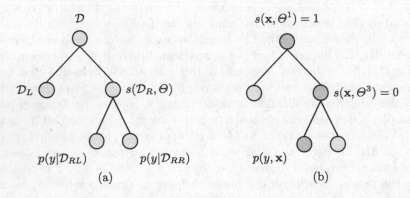

Fig. 7. Random Forests: (a) The tree is build recursively splitting the training data $\mathcal{D}$ and finally estimating a model $p(y|\mathcal{D}_*)$ for each leaf node. (b) During inference a sample $\mathbf{x}$ is traversed down according to the learned splitting functions s (i.e., the parameters Θ^*) the tree and finally classified based on the model of the leaf node.

A splitting function $s(\mathbf{x}, \Theta)$ is typically parameterized by two values: (i) a feature dimension $\Theta_1 \in \{1, \dots, N\}$ and (ii) a threshold $\Theta_2 \in \mathbb{R}$. The splitting function is then defined as

$$s(\mathbf{x}, \Theta) = \begin{cases} 0 & \text{if } \mathbf{x}(\Theta_1) < \Theta_2 \\ 1 & \text{otherwise} \end{cases}, \tag{18}$$

where the outcome defines to which child node the sample $\mathbf{x}$ is routed.

Each node i chooses the best splitting function Θ^i out of a randomly sampled set by optimizing the information gain

$$\Delta(\Theta^i) = \frac{|\mathcal{D}_L|}{|\mathcal{D}_L| + |\mathcal{D}_R|} H(\mathcal{D}_L) + \frac{|\mathcal{D}_R|}{|\mathcal{D}_L| + |\mathcal{D}_R|} H(\mathcal{D}_R), \tag{19}$$

where $\mathcal{D}_L$ and $\mathcal{D}_R$ are the sets of data samples that are routed to the left and right child nodes, according to $s(\mathbf{x}, \Theta^i)$; $H(\mathcal{D})$ is the local score of a set $\mathcal{D}$ of data samples, which can either be the negative entropy

$$H(\mathcal{D}) = -\sum_{c=1}^{C} [p(c|\mathcal{D}) \cdot log(p(c|\mathcal{D}))], \tag{20}$$

where C is the number of classes, and $p(c|S)$ is the probability for class c, estimated from the set S, or the Gini Index [46].

The most important application of RFs in medical image analysis is the automatic segmentation of cells, organs or tumors, typically building on a multi-class classification forests. For example, in [48] an approach for segmenting high-grade gliomas and their sub-regions from multi-channel MR images is presented. By using context-aware features and the integration of a generative model of tissue appearance only little pre-processing and no explicit regularization is required, making the approach computationally very efficient. A different approach was presented in [49], where a joint classification-regression forest was trained, that captures both structural and class information. In this way, not only a class label is predicted but also the distance to the object boundary. Applied on 3-dimensional CT scans the final task of multi-organ segmentation can be solved very efficiently.

Related to the previous task is the application of detecting and localizing anatomy. For example, [50] introduces an approach for localizing vertebras using a combined segmentation and localization approach. For this purpose a RF is trained using features images obtained form a standard filter bank, where the output is then used – together with the original image – to generate candidate segmentations for each class, which are finally weighted. In contrast, [51] addresses the problem of localizing organs such as spleen, liver or heart. To this end, visual features are extracted from the imaging data and a regression forest is trained, allowing for a direct mapping form voxels to organ locations and size. In particular, the approach deals with both magnetic resonance (MR) and computer tomography (CT) images, also showing the generality and flexibility of RFs. A similar approach is addressed in [52], also estimating local landmark points finally paving the way for automatic age estimation [53].

For a detailed overview on Random Forests we would like to refer to [47], where a deep theoretical discussion as well as an overview of different applications in the field of medical image analysis are given.

Deep Learning

Event though the main ideas of neural networks are dating back to the 1940's (i.e., [54,55]), they just become recently very popular due the success of convolutional neural networks [56,57]. In general, neural networks, are biologically inspired and can be described as a directed graph, where the nodes are related to neurons/units and the edges describe the links between them.

As illustrated in Fig. 8, each unit j receives a weighted sum of inputs a_i of connected units i, where the weights $w_{i,j}$ determine the importance of the connection. To estimate the output a_j this linear combination is then fed into a so called activation function. More formally, the output a_j is estimated as follows:

$$a_j = g\left(\sum_{i=0}^{n} w_{i,j} a_i\right). \tag{21}$$

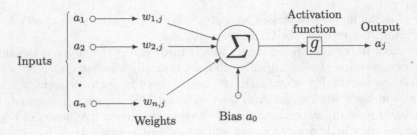

Fig. 8. The general model of a single neuron: the weighted inputs a_i are summed up and fed into an activation function $g(\cdot)$ yielding the output a_j.

Popular choices for the activation function which are widely used are

- Linear function: $g(x) = x$,
- Hard threshold function: $g(x) = \begin{cases} 1 & \text{if } x > \frac{1}{2} \\ 0 & \text{otherwise,} \end{cases}$
- Sigmoid function: $g(x) = \frac{1}{1+e^{-x}}$,
- Hyperbolic tangent function: $g(x) = tanh(x)$,
- Rectified Linear Units (ReLU): $g(x) = max(0, x)$.

In general, we can distinct two different kinds of networks. First, feed-forward networks can be described as acyclic graphs, having connections only in one direction. In this way, the network describes a function of the input. In contrast, recurrent networks (RNNs) can be considered graphs with loops, as receiving their outputs again as input (thus being non-linear systems). In this way, an internal state (short term memory) can be described. Thus, RNNs are widely used in applications such as speech recognition or in activity recognition [58], whereas in image processing mainly feed-forward networks are of relevance [59].

Neural networks are typically arranged in layers V_i consisting of single units as described above, such that each unit receives input only from units from the previous layer, where $|V| = T$ the depth of the network. V_0 is referred to as the input layer, V_T as the output layer, and $V_1, \ldots, V_{T-1}$ are called the hidden layers. A simple example of such a network with two hidden layers is illustrated in Fig. 9. When dealing with multiple hidden layers, we talk about deep learning. In general, this allows to learn complex functions, where different layers cover different kind of information. For example, in object detection a first layer may describe oriented gradients, a second layer some kind of edges, a third layer would assemble those edges to object descriptions, where a subsequent layer would describe the actual detection task. This example also illustrates an important property of deep learning: we can learn feature representations and do not need to design features by hand!

In general, the goal of supervised learning is to modify the model parameters such that subsequently the error of an objective function is reduced. For neural networks, this is typically solved via the stochastic gradient descend (SGD) approach. The main idea is to repeatedly compute the errors for many small sets and

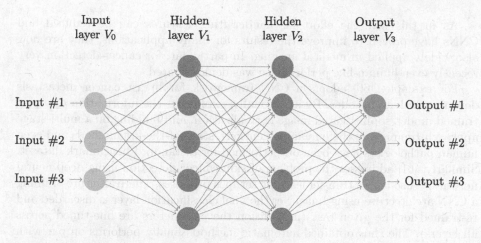

| Input | Hidden | Hidden | Output |
| layer V_0 | layer V_1 | layer V_2 | layer V_3 |

Input #1 →

Input #2 →

Input #3 →

→ Output #1

→ Output #2

→ Output #3

Fig. 9. Deep feed-forward neural network with two hidden layers (blue balls). In addition, the input layer (green balls) and the output layer (red points) are illustrated. (Color figure online)

to adjust the model according to a averaged response. Thus, the name can be explained as a gradient method – typically using the back-propagation approach – is used and the computation based on small sets of samples is naturally noisy.

The most prominent and most successful deep learning architecture are Convolutional Neural Networks (CNN), why these terms are often used interchangeable. Even though naturally inspired by image processing problems the same ideas can also be beneficial for other tasks. One key aspect of CNNs is that the are structured in a series of different layers: convolutional layers, pooling layers, and fully connected layers. Convolutional layers can be considered feature maps, where each feature map is connected to local patches in the feature map in the previous layer. In contrast, pooling layers merge similar features into one (i.e., relative positions of features might vary in the local range). Typical, several stages of convolutional and pooling layers are stacked together. Finally, there are fully connected layers generating the output of the actual task. A typical architecture for such a CNN is illustrated in Fig. 10.

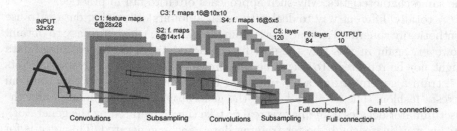

Fig. 10. Typical convolutional neural network: LeNet-5 [56].

As in this way the effort for handcrafting features can be reduced and CNNs have proven to improve the results for many applications, they are now also widely applied in medical imaging. In particular, for cancer detection very recently even human-like performance was demonstrated.

For example, [60] adopts a CNN framework for breast cancer metastasis detection in lymph nodes. By additionally, exploiting the information of a pre-trained model, sophisticated image normalization, and building on a multi-stage approach (mimicking the human perception), state-of-the-art methods and even human pathologists have been outperformed for a standard benchmark dataset. Similarly, [61] addresses the problem of skin cancer detection using deep neural networks. Also here a transfer learning setup is proposes, where after pre-training a CNN architecture using ImageNet the final classification layer is discarded and re-trained for the given task (in addition the parameters are fine-tuned across all layers). The thus obtained automatic methods finally performs on par with human pathologist on different task.

Even though this demonstrates, that Deep Learning could be very beneficial in the medical domain, the main challenge is to cope with the problem that often the rather large amount of required training data is not available. Thus, there has been a considerable interest in approaches that can learn from a small number of training samples. The most common and straight forward way is to use data augmentation, where additional training samples are generated via variation of the given data: rotation, elastic deformation, adding noise, etc. One prominent example for such approaches is U-Net [62], which demonstrated that for biomedical image segmentation state-of-the-art results can be obtained, even when the model was trained just from a few samples.

Even though this simple approach often yields good results, it is limited as only limited variations can be generated from the given data. A different direction is thus to build on ideas from transfer learning [63]. The key idea is to pre-train a network on large publicly available datasets and then to fine-tune it for the given task. For example, [64] fine-tunes the VGG-16 network, which is already pre-trained using a huge amount of natural images, to finally segment pancreas from MR images. In addition, a CRF step is added for the final segmentation. Another way would be to use specific prior knowledge about the actual task [65]. However, this information is often not available and, as mentioned above, medical image data and natural images are often not sharing the same characteristics, why such approaches often to fail in practice.

A totally different way to deal with small amounts of training data is to use synthetically generated samples for training (e.g., [66]), which are easy to obtain. However, again in this way the specific characteristic of the given image data might not be reflected. To overcome this problem, Generative Adversarial Nets [67] train a generator and a discriminator framework in a competitive Random Forests fashion. The key idea is that the generator synthesizes images and the discriminator decides if an image is real or fake (i.e., generated by the generator). In this way, increasingly better training data can be generated. This idea is for example exploited by [68] to better model the nonlinear relationship between CR and MR images.

Further Reading: For a short review on Deep Learning we would like to refer to [13], a detailed review of related work can be found in [69], and a very detailed technical overview ins given in [70].

Summary

As demonstrated in this section there are several ways to address the wide range of applications in medical imaging. Even though there exist special approaches for specific application, we focused on three versatile and thus widely used approaches, namely, Support Vector Machines (SVMs), Random Forests (RFs), and Deep Learning (DL). Where SVMs are general working horses for different applications, RFs have demonstrated to cope with the particular characteristics of medical imaging data very well. However, for both approaches well-engineered handcrafted features are necessary, which are often hard to define and compute. This problem can be overcome by using DL approaches, as the required features can be learned implicitly in an end-to-end manner. However, the main drawback of such approaches is that they require a huge amount of training data to yield competitive results, which is often not available in practice. There are several approaches which help to moderate this problem, but in general dealing with a small data is still a big problem. Thus, still other methods such as SVM and RF are valid choices for medical imaging problems. In addition, a key aspect that is often neglected is that there are often good biases available, either defined by the specific task or by available human experts, which are not considered (sufficiently) up to now!

5   Secure Cooperation with Experts

Securing the data life cycle is a problem that just recently gained a lot of additional attention, mainly due to the General Data Protection Regulation (GDPR)[1] that not only established baselines for securing sensitive information throughout the European Union, but also increases the penalties to be applied in case of violation. This regulation will come into effect in May 2018 either directly or by adoption into national law. It is concerned with all major issues regarding the processing of personal sensitive information, most notably it deals with the legal requirements regarding data collection, consent regarding processing, anonymization/pseudonymization, data storage, transparency and deletion [71]. Still, the major issue here is that many details are currently not defined, e.g. whether deletion needs to be done on a physical or simply logical level, or how strong the anonymization-factors need to be [72]. Furthermore, some parts are formulated in a way that cannot be achieved with current technological means, e.g. de-anonymization being impossible in any case, as well as the antagonism

---

[1] Regulation (EU) 2016/679 of the European Parliament and of the council of 27 April 2016 on the protection of natural persons with regard to the processing of personal data and on the free movement of such data, and repealing directive 95/46/EC (General Data Protection Regulation).

between deletion and transparency. Thus, the issue of securing sensitive information is one of the really big challenges in machine learning in health related environments [73].

5.1 Data Leak Detection

While many of the issues outlined above seem not that relevant for pathological data at first glance, other questions regarding data security still prevail: Data is considered to be the new oil, meaning that data itself constitutes a significant value. One of the major issues in machine learning based research lies in the issue of cooperation between data owners and other entities. With ever new techniques arriving on the scene requiring the cooperation of various experts in the areas of modeling, machine learning and medicine, data needs to be shared between different entities, often working at different institutions. This opens up the issue of detecting data misuse, especially the unsolicited dissemination of data sets.

Measures against data leakage can be divided into two major categories, those protecting the data from being leaked (proactive measures) and those enabling the detection and attribution of data leaks (reactive measures). Proactive measures typically include limitations on the data exchange:

- Sealed research environments like dedicated servers that run all the analysis software and contain all the data, without export possibilities, as well as sporting mechanisms for controlling, which researcher utilized which information set. While this does not constitute a 100 percent protection against malicious experts trying to extract information, in does help against accidental data loss.
- Aggregating the data as much as possible can be a solution too for reducing the amount of sensitive information that an expert is given access to. Still, aggregation often introduces a significant error and can render the data practically worthless for the analysis.
- Oracle-style measures like differential privacy [74] do not deliver the whole data set to the experts but rather require the expert to choose the type of analysis he/she wants to run on the data without seeing the data sets. Typically, this is done via issuing "Select"-statements that are run against the database. Measures like differential privacy introduce distortion for data protection purposes, as well as limit the amount of information the expert can retrieve from the data.

While these proactive measures certainly do have their merits, they often pose a serious obstacle to cooperation with the world-best experts in a field, either due to geographical issues, or simply because the data is not fine-grained enough to utilize the whole potential of the information inside the data sets. Reactive approaches aim at identifying data leaks instead of preventing exchange, i.e. the data is distributed to the partners in a form suitable for analysis (while, of course, still considering issues of data protection like required anonymization), but it is marked in order to make each data set unique for each partner it is distributed to. These marks are typically called *fingerprints* and are required to possess the following features [75]:

- The fingerprinting mechanism must be capable to uniquely identify a user.
- It must not be possible for a user to identify the mark and subsequently remove it (without drastically reducing the utility of the data).
- Even for a combination of attackers, the fingerprint must be resilient against inference attacks, i.e. it must not be possible to calculate the fingerprinting marks, even when given several differently marked versions of the same data set.
- No wrongful accusations must be possible, even in case several attacks work together in order to change the fingerprint to put blame on an uninvolved partner.
- The fingerprint must be tolerant against a certain amount of errors introduced by the users, i.e. it must not get useless in case the attackers change some portion of the marked data.

Furthermore, depending on the actual form of the data, there are typically some additional issues that require consideration, e.g., how stable the fingerprint needs to be in case only part of the data is leaked (e.g. half a picture, some records from a sample). Since the GDPR often requires the anonymization of sensitive information, one interesting approach lies in the development of combined methods that use intrinsic features of the anonymization technique in order to generate a selection of different data sets that can be uniquely identified. In the past, such a fingerprinting approach was proposed for structured data in tables [76], still, the number of different fingerprints that can be assigned to the same data set while providing resilience against collaborating attackers is rather low and mainly depends on the actual data, especially when obeying the requirement for resilience against colluding attackers as outlined above [77].

5.2   Deletion of Data

Typically, the deletion of data is not considered to be a major problem in most applications, as it is mostly a matter of freeing resources. Still, against the background of the GDPR, this topic becomes increasingly important to consider, especially, since it is extremely complicated to delete information in modern, complex environments [72]. Databases are a major example, why deletion of information can be a major problem: ACID-compliance [78] is a major requirement of modern database management systems and requires the database product to ensure the atomicity of certain operations, i.e. operations are either carried out as a whole, or not at all, always leaving the database to be in a consistent state, even in case of a server crash. Furthermore, mechanisms for undoing operations, most notable rollback mechanisms, are currently state-of-the-art and expected by database users. Thus, a lot of additional information is required to be stored in various internal mechanisms of the database, e.g. the transaction mechanism, which is responsible for collecting all operations changing the database and enables rollbacks and crash-recovery.

Still, in typical database environments, even simple removal of data from the database itself is non-trivial, considering the way data is stored: The records inside a table are stored inside a tree-structure, more notably a B^+-tree [79]:

– For the number m_i of elements of node i holds $\frac{d}{2} \le i \le d$, given a pre-defined d for the whole tree, the *order* of the tree. The ony example of this rule is the root r with $0 \le r \le d$.
– Each non-leaf-node with m elements possesses $m+1$ child nodes, $\frac{d}{2} \le m \le d$.
– The tree is balanced, i.e. all leaf nodes are on the same level.
– In contrast to the normal B-tree, the inner nodes of the B^+-tree solely store information required for navigating through the tree, the actual data is stored in the leaf nodes, making the set of leafs forming a partition of the whole data set.
– The elements inside the nodes are stored as sorted lists.

In databases like MySQL, the (mandatory) *primary key* of each table is used to physically organize the data inside the table in the form of a B^+-Tree, the secondary indices are merely search-trees of their own, solely containing links to the tree built by the primary key.

When data is deleted from the indexing tree built by the primary key, the database searches for the leaf node containing the required element. Since databases are built in order to facilitate fast operations, the data inside the leaf node is not overwritten, but simple unlinked from the sorted list inside said node.

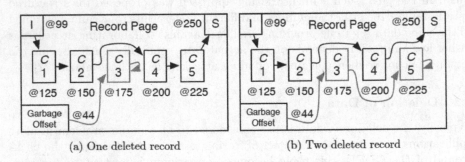

(a) One deleted record   (b) Two deleted record

Fig. 11. Deletion in MySQL [80].

Figure 11 gives a short overview on the deletion process: The record in question is unlinked from the linked list and added to the so-called *garbage collection*, which marks the space of the record as free for storing new data. Still, the data is not technically overwritten at the point of deletion and can be reconstructed quite simple, as long as the space has not been used again, which depending on the usage patterns of the database, is unpredictable and might take a long time.

Still, even the actual form of the search tree itself might yield information on data already deleted from the table [81]: Let B be a B^+-tree with $n > b$ elements which are added in ascending order. Then it holds true that the partition of the leafs of B has the following structure:

$$n = \sum_{i=1}^{k} a_i, \text{ with } a_i = \frac{b}{2} + 1, \forall i \neq k \text{ and } a_k \geq \frac{b}{2}.$$

While this theorem allows only very limited detection for practical purposes under certain circumstances, especially due to database internal reorganization mechanisms destroying the structure, there are instances where information can be retrieved from the structure of the B^+-tree.

As [72] outlined, there is currently no definition in the GDPR, how data must be deleted, i.e. whether this needs to be done physically, or only logically. Still, when looking at earlier national legal counterparts concerning data protection, several legislations (e.g. in Austria) used a very strict interpretation of the term "deletion", i.e. physical deletion. In addition, the GDPR deals a lot in absolutes, either jurisdiction is required to relax these absolutes, or new technical means for removing evidence deeply embedded in complex systems are required.

5.3 Information Extraction

We have already seen how databases can be secured via differential privacy and other query mechanisms, however most data in a clinical environment exist in the form of Electronic Health records whose entries are mostly unstructured free-text documents. As there is no guaranteed way to anonymize unstructured data, we first need to extract identified bits of information and convey them to a more organized data structure.

Information Extraction (IE) is the art of finding relevant bits of specific meaning within unstructured data - this can be done either via (1) low-level IE - usually by means of dictionary or RegExp based approaches [82,83] which utilize extensive corpora of biomedical vocabulary and are readily available in libraries such as Apache cTakes; the disadvantage of most standard solutions is the lack of their ability to correctly identify context, ambiguous, synonymous, polysemous or even just compound words; or (2) higher level IE, usually in the form of a custom-built *natural language processing (NLP)* pipelines. Often, the purpose of such pipelines is *Named Entity Recognition (NER)*, which is the task of labeling terms of specific classes of interest, like People, Locations, or Organizations.

It was noted [84] that NER is more difficult in specialized fields, as terms have more narrow meanings (abbreviations can mean different things, e.g.). The authors of [85] describe NER as a sequence segmentation problem to which they apply *Conditional Random Fields (CRF)*, a form of undirected statistical graphical models with Markov independence assumption, allowing them to extract orthographic as well as semantic features. More recently, even Neural Networks have been utilized for NER [86] with performance at state-of-the-art levels, partly incorporating a new form of concept space representations for terms called embeddings, which use a form of dimensionality reduction to compress vocabulary-sized feature vectors into (mostly 50-300 dimensional) concept vectors [87].

5.4   Anonymization

After having condensed all available material into labeled information, we can filter them through formal anonymization approaches, of which k-anonymity [88] stands as the most prominent. K-anonymity requires that a release of data shall be clustered into equivalence groups of size $>= k$ in which all quasi identifiers (non-directly identifying attributes such as age, race or ZIP code) have been generalized into duplicates; generalization itself can be pictured as an abstraction of some information to a more inclusive form, e.g. abstracting a ZIP code of *81447* to textit81***, thereby being able to potentially cluster it with all other ZIP codes starting with *81****.

Beyond k-anonymity exist refinements such as l-diversity [89], t-closeness [90] and δ-presence [91] for purely tabular data, as well as a number of individual methods for social network anonymization [92, 93]. They all operate on the concept of structured, textual quasi identifiers and imply a trade-off between data utility and privacy of a data release - a higher degree of anonymization provides more security against identifying a person contained in the dataset, but reduces the amount of usable information for further studies or public statistics.

This leads us to the field of *Privacy aware Machine Learning (PaML)* which is the application of ML techniques to anonymized (or in any way perturbed) datasets. Obviously one cannot expect the performance of such algorithms to equal their counterparts executed on the original data [94], instead the challenge is to produce anonymized datasets which yield results of similar quality than the original. This can be achieved by cleverly exploiting statistical properties of such data, e.g. outliers in the original might affect ML performance as well as induce higher levels of generalization necessary to achieve a certain factor of k; by first removing those outliers an anonymized version can actually retain enough information to rival its unperturbed predecessor [95].

5.5   Image Data Integration

For medical purposes, images have long been considered quasi-identifiers [96, 97], as one can easily picture faces allowing a relatively exact reconstruction of a persons identity (depending on the quality of algorithms used). In the case of pathological images containing multiple features and feature groups in relation to one another, any subset of such information could conceivably be combined with a patient's EHR, thus enabling re-identification. On the other hand, selected features also complement a patients EHR and therefore provide a more complete overview of the patient's condition facilitating higher precision in diagnosis, especially in cases when doctors overlook or forget to record symptoms. In approximating an answer to the question'how can one anonymize images', we would like to provide a simple (therefore unrealistic) illustration (Fig. 12), in which the bottom row depicts 8 original face images, whereas the subsequent vertical rows represent progressive morphings of pairs of samples below, arriving at a most general male/female hybrid at the top. In a realistic clinical setting, a useful effect could be achieved by learning features from individual samples (faces,

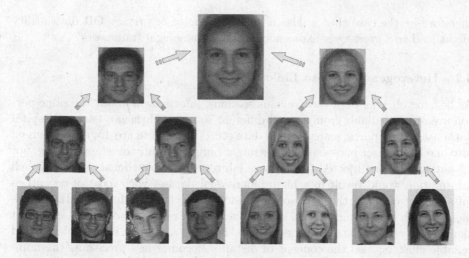

Fig. 12. Depiction of a possible (naive) face generalization hierarchy by simple morphing of aligned images. Realistically, one would first scan images for salient features, then cluster faces via feature similarity, subsequently morphing the generalized features back into a pictograph or artificially generated face.

pathologic slides etc.), clustering those traits by similarity and then merging them together into collective representations of groups.

5.6 *Omics Data Integration

In contrast to images it seems very doubtful if *omics-data can be perturbed in a meaningful way to protect a person's identity. After all, genes, proteins etc. are building blocks of larger structures, and changing even one gene to a variant form (called an *allele*) can have significant repercussions on an organism's phenotype. So in the field of *omics research, the issue of privacy is treated a little differently: Given e.g. a GWAS (genome-wide association study) and the genetic profile of an individual person, the question arises with what certainty a classifier could determine if that person participated in said study. This entails the need to perturb the results of a study - a distribution of measurements, like allele frequencies [98] - rather than a database of personal information, which lends itself ideally to the already described mechanism of ϵ-differential privacy. The authors of [99] even tailored the method to GWAS study data in case of the presence of population stratification and studied its effect on the output of the EIGENSTRAT and LMM (Linear Mixed Model) algorithms typically used on a rheumatoid arthritis GWAS dataset. To what extent those methods can actually protect people from identification is a point of open discussion: while some researchers [100] claim that even with standard statistical methods a binary classification result (AUC) of reasonably close to 1 can be achieved, others [101] point out that DNA matching in itself is not equivalent to de-identification and even if possible, would take tremendous time and computational power. It might therefore be the case that a false notion of a choice of privacy OR data utility might lead to a gross over-expansion of the privacy legal framework.

5.7 Heterogeneous Data Linkage

As became obvious from the previous sections, information about the same person are often available from several different sources (physician letters, hospital databases, lab reports, scans, *omics data etc.). These data are not easily merged into one big dataset because coverage might only be slightly overlapping (e.g. not all patients were subjected to the same lab tests). Simple concatenation of such information would result in a high-dimensional dataset with most of its entries missing, introducing the curse-of-dimensionality when conducting ML experiments. With increasing dimensionality, the volume of the space increases so fast that the available data becomes sparse, hence it becomes impossible to find reliable clusters; also the concept of distance becomes less precise as the number of dimensions grows, since the distance between any two points in a given data set converges; moreover, different clusters might be found in different sub spaces, so a global filtering of attributes is also not sufficient. A solution might be found in graph-based representations of such data, where node types can represent patients or different forms of examinations, resources, etc.; in the case of anonymizing, we not only have to generalize node information but also consider neighborhood structure which could provide an adversary with additional hints for attack vectors. Apart from dealing with graph anonymization, which is also a hard problem [102], an interesting challenge lies in describing parameters of the underlying stochastic process precisely enough so one can re-populate a graph from its anonymized form; this generatively perturbed graph should on the one hand meet privacy requirements, yet allow scientists to conduct ML experiments yielding satisfactory performance.

6 Future Challenges

Much future research has to be done, particularly in the fields of Multi-Task Learning and Transfer Learning to go towards Multi-Agent-Hybrid Systems as applications of the iML-approach.

6.1 Future Challenge 1: Multi-task Learning

Multi-task learning (MTL) aims to improve the prediction performance by learning a problem together with multiple, different but related other problems through shared parameters or a shared representation. The underlying principle is *bias learning* based on *Probably Approximately Correct* learning (PAC learning) [19]. To find such a bias is still the hardest problem in any ML task and essential for the initial choice of an appropriate hypothesis space, which must be large enough to contain a solution, and small enough to ensure a good generalization from a small number of data sets. Existing methods of bias generally require the input of a human-expert-in-the-loop in the form of heuristics and domain knowledge to ensure the selection of an appropriate set of features, as such features are key to learning and understanding. However, such methods are limited by the accuracy and reliability of the expert's knowledge (robustness of the human) and also by the extent to which that knowledge can be transferred to new tasks (see next subsection). Baxter (2000) [103] introduced a model of bias learning which builds on the PAC learning model which concludes that learning multiple related tasks reduces the sampling burden required for good generalization. A bias which is learnt on sufficiently many training tasks is likely to be good for learning novel tasks drawn from the same environment (the problem of transfer learning to new environments is discussed in the next subsection). A practical example is *regularized MTL* [104], which is based on the minimization of regularization functionals similar to Support Vector Machines (SVMs), that have been successfully used in the past for singletask learning. The regularized MTL approach allows to model the relation between tasks in terms of a novel kernel function that uses a taskcoupling parameter and largely outperforms singletask learning using SVMs. However, multi-task SVMs are inherently restricted by the fact that SVMs require each class to be addressed explicitly with its own weight vector. In a multi-task setting this requires the different learning tasks to share the *same set of classes.* An alternative formulation for MTL is an extension of the large margin nearest neighbor algorithm (LMNN) [105]. Instead of relying on separating hyper-planes, its decision function is based on the nearest neighbor rule which inherently extends to many classes and becomes a natural fit for MTL. This approach outperforms state-of-the-art MTL classifiers, however, much open research challenges remain open in this area [106].

6.2   Future Challenge 2: Transfer Learning

A huge problem in ML is the phenomenon of *catastrophic forgetting*, i.e. when a ML algorithm completely and abruptly "forgets" how to perform a learned task once transferred to a different task. This is a well-known problem which affects ML-systems and was first described in the context of connectionist networks [107]; whereas natural cognitive systems rarely completely disrupt or erase previously learned information, i.e. natural cognitive systems do not forget "catastrophically" [108]. Consequently the challenge is to discover how to avoid the problem of catastrophic forgetting, which is a current hot topic [109].

According to Pan and Yang (2010) [21] a major requirement for many ML algorithms is that both the training data and future (unknown) data must be in the same feature space and show similar distribution. In many real-world applications, particularly in the health domain, this is not the case: Sometimes we have a classification task in one domain of interest, but we only have sufficient training data in another domain of interest, where the latter data may be in a completely different feature space or follows a different data distribution. In such cases transfer learning would greatly improve the performance of learning by avoiding much expensive data-labeling efforts, however, much open questions remain for future research [110].

6.3   Future Challenge 3: Multi-Agent-Hybrid Systems

Multi-Agent-Systems (MAS) are collections of many agents interacting with each other. They can either share a common goal (for example an ant colony, bird flock, or fish swarm etc.), or they can pursue their own interests (for example as in an open-market economy). MAS can be traditionally characterized by the facts that (a) each agent has incomplete information and/or capabilities for solving a problem, (b) agents are autonomous, so there is no global system control; (c) data is decentralized; and (d) computation is asynchronous [111]. For the health domain of particular interest is the *consensus problem*, which formed the foundation for distributed computing [112]. The roots are in the study of (human) experts in group consensus problems: Consider a group of humans who must act together as a team and each individual has a subjective probability distribution for the unknown value of some parameter; a model which describes how the group reaches agreement by pooling their individual opinions was described by DeGroot [113] and was used decades later for the aggregation of information with uncertainty obtained from multiple sensors [114] and medical experts [115]. On this basis Olfati-Saber et al. [116] presented a theoretical framework for analysis of consensus algorithms for networked multi-agent systems with fixed or dynamic topology and directed information flow. In complex real-world problems, e.g. for the epidemiological and ecological analysis of infectious diseases, standard models based on differential equations very rapidly become unmanageable due to too many parameters, and here MAS can also be very helpful [117]. Moreover, collaborative multi-agent reinforcement learning has a lot of research potential for machine learning [118].

7   Conclusion

Machine learning for digital pathology poses a lot of challenges, but the premises are great. An autonomous pathologist, acting as digital companion to augment real pathologists can enable disruptive changes in future pathology and in whole medicine. To reach such a goal much further research is necessary in collecting, transforming and curating explicit knowledge, e.g. clinical data, molecular data and e.g. lifestyle information used in medical decision-making.

Digital Pathology will highly benefit from interactive Machine Learning (iML) with a pathologist in the loop. Currently, modern deep learning models are often considered to be "black-boxes" lacking explicit declarative knowledge representation. Even if we understand the mathematical theories behind the machine model it is still complicated to get insight into the internal working of that model, hence black box models are lacking transparency and the immediate question arises: "Can we trust our results?" In fact: "Can we explain how and why a result was achieved?" A classic example is the question "Which objects are similar?", but an even more interesting question is "Why are those objects similar?". Consequently, in the future there will be urgent demand in machine learning approaches, which are not only well performing, but transparent, interpretable and trustworthy. If human intelligence is complemented by machine learning and at least in some cases even overruled, humans must be able to understand, and most of all to be able to interactively influence the machine decision process. A huge motivation for this approach are rising legal and privacy aspects, e.g. with the new European General Data Protection Regulation (GDPR and ISO/IEC 27001) entering into force on May, 25, 2018, will make black-box approaches difficult to use in business, because they are not able to explain why a decision has been made.

This will stimulate research in this area with the goal of making decisions interpretable, comprehensible and reproducible. On the example of digital pathology this is not only useful for machine learning research, and for clinical decision making, but at the same time a big asset for the training of medical students. Explainability will become immensely important in the future.

Acknowledgments. We are grateful for valuable comments from the international reviewers.

References

1. Madabhushi, A., Lee, G.: Image analysis and machine learning in digital pathology: Challenges and opportunities. Med. Image Anal. **33**, 170–175 (2016)
2. Holzinger, A.: Trends in interactive knowledge discovery for personalized medicine: Cognitive science meets machine learning. IEEE Intell. Inf. Bull. **15**, 6–14 (2014)
3. Su, X., Kang, J., Fan, J., Levine, R.A., Yan, X.: Facilitating score and causal inference trees for large observational studies. J. Mach. Learn. Res. **13**, 2955–2994 (2012)
4. Huppertz, B., Holzinger, A.: Biobanks – a source of large biological data sets: open problems and future challenges. In: Holzinger, A., Jurisica, I. (eds.) Interactive Knowledge Discovery and Data Mining in Biomedical Informatics. LNCS, vol. 8401, pp. 317–330. Springer, Heidelberg (2014). doi:10.1007/978-3-662-43968-5_18
5. Mattmann, C.A.: Computing: A vision for data science. Nature **493**, 473–475 (2013)
6. Otasek, D., Pastrello, C., Holzinger, A., Jurisica, I.: Visual data mining: effective exploration of the biological universe. In: Holzinger, A., Jurisica, I. (eds.) Interactive Knowledge Discovery and Data Mining in Biomedical Informatics. LNCS, vol. 8401, pp. 19–33. Springer, Heidelberg (2014). doi:10.1007/978-3-662-43968-5_2
7. Ghahramani, Z.: Probabilistic machine learning and artificial intelligence. Nature **521**, 452–459 (2015)
8. Teh, Y.W., Jordan, M.I., Beal, M.J., Blei, D.M.: Hierarchical dirichlet processes. J. Am. Stat. Assoc. **101**, 1566–1581 (2006)
9. Houlsby, N., Huszar, F., Ghahramani, Z., Hernndez-lobato, J.M.: Collaborative gaussian processes for preference learning. In: Pereira, F., Burges, C., Bottou, L., Weinberger, K. (eds.) Advances in Neural Information Processing Systems (NIPS 2012), pp. 2096–2104 (2012)
10. Holzinger, A.: Introduction to machine learning and knowledge extraction (make). Mach. Learn. Knowl. Extr. **1**, 1–20 (2017)

11. Shahriari, B., Swersky, K., Wang, Z., Adams, R.P., de Freitas, N.: Taking the human out of the loop: A review of bayesian optimization. Proc. IEEE **104**, 148–175 (2016)

12. Kim, W., Choi, B.J., Hong, E.K., Kim, S.K., Lee, D.: A taxonomy of dirty data. Data Min. Knowl. Disc. **7**, 81–99 (2003)

13. LeCun, Y., Bengio, Y., Hinton, G.: Deep learning. Nature **521**, 436–444 (2015)

14. Lee, S., Holzinger, A.: Knowledge discovery from complex high dimensional data. In: Michaelis, S., Piatkowski, N., Stolpe, M. (eds.) Solving Large Scale Learning Tasks. Challenges and Algorithms. LNCS (LNAI), vol. 9580, pp. 148–167. Springer, Cham (2016). doi:10.1007/978-3-319-41706-6_7

15. Holzinger, A., Plass, M., Holzinger, K., Crisan, G.C., Pintea, C.M., Palade, V.: A glass-box interactive machine learning approach for solving NP-hard problems with the human-in-the-loop (2017). arXiv:1708.01104

16. Holzinger, A.: Interactive machine learning for health informatics: When do we need the human-in-the-loop? Brain Informatics (BRIN) **3** (2016)

17. Holzinger, A., Plass, M., Holzinger, K., Crişan, G.C., Pintea, C.-M., Palade, V.: Towards interactive machine learning (iML): applying ant colony algorithms to solve the traveling salesman problem with the human-in-the-loop approach. In: Buccafurri, F., Holzinger, A., Kieseberg, P., Tjoa, A.M., Weippl, E. (eds.) CD-ARES 2016. LNCS, vol. 9817, pp. 81–95. Springer, Cham (2016). doi:10.1007/978-3-319-45507-5_6

18. Fayyad, U., Piatetsky-Shapiro, G., Smyth, P.: From data mining to knowledge discovery in databases. AI Mag. **17**, 37–54 (1996)

19. Valiant, L.G.: A theory of the learnable. Commun. ACM **27**, 1134–1142 (1984)

20. Holzinger, A.: On topological data mining. In: Holzinger, A., Jurisica, I. (eds.) Interactive Knowledge Discovery and Data Mining in Biomedical Informatics. LNCS, vol. 8401, pp. 331–356. Springer, Heidelberg (2014). doi:10.1007/978-3-662-43968-5_19

21. Pan, S.J., Yang, Q.: A survey on transfer learning. IEEE Trans. Knowl. Data Eng. **22**, 1345–1359 (2010)

22. Demichelis, F., Barbareschi, M., Dalla Palma, P., Forti, S.: The virtual case: a new method to completely digitize cytological and histological slides. Virchows Arch. **441**, 159–161 (2002)

23. Bloice, M., Simonic, K.M., Holzinger, A.: On the usage of health records for the design of virtual patients: a systematic review. BMC Med. Inform. Decis. Mak. **13**, 103 (2013)

24. Turkay, C., Jeanquartier, F., Holzinger, A., Hauser, H.: On computationally-enhanced visual analysis of heterogeneous data and its application in biomedical informatics. In: Holzinger, A., Jurisica, I. (eds.) Interactive Knowledge Discovery and Data Mining in Biomedical Informatics. LNCS, vol. 8401, pp. 117–140. Springer, Heidelberg (2014). doi:10.1007/978-3-662-43968-5_7

25. Ferreira, R., Moon, B., Humphries, J., Sussman, A., Saltz, J., Miller, R., Demarzo, A.: The virtual microscope. In: Proceedings of the AMIA Annual Fall Symposium, pp. 449–453 (1997)

26. Barbareschi, M., Demichelis, F., Forti, S., Palma, P.D.: Digital pathology: Science fiction? Int. J. Surg. Pathol. **8**, 261–263 (2000). PMID: 11494001

27. Hamilton, P.W., Wang, Y., McCullough, S.J.: Virtual microscopy and digital pathology in training and education. Apmis **120**, 305–315 (2012)

28. Dandu, R.: Storage media for computers in radiology. Indian J. Radiol. Imag. **18**, 287 (2008)

29. Reeder, M.M., Felson, B.: Gamuts in Radiology: Comprehensive Lists of Roentgen Differential Diagnosis. Pergamon Press (1977)
30. Goolsby, A.W., Olsen, L., McGinnis, M., Grossmann, C.: Clincial data as the basic staple of health learning - Creating and Protecting a Public Good. National Institute of Health (2010)
31. McDermott, J.E., Wang, J., Mitchell, H., Webb-Robertson, B.J., Hafen, R., Ramey, J., Rodland, K.D.: Challenges in biomarker discovery: combining expert insights with statistical analysis of complex omics data. Expert Opinion Med. Diagn. **7**, 37–51 (2013)
32. Swan, A.L., Mobasheri, A., Allaway, D., Liddell, S., Bacardit, J.: Application of machine learning to proteomics data: Classification and biomarker identification in postgenomics biology. Omics-a J. Integr. Biol. **17**, 595–610 (2013)
33. Jeanquartier, F., Jean-Quartier, C., Schreck, T., Cemernek, D., Holzinger, A.: Integrating open data on cancer in support to tumor growth analysis. In: Renda, M.E., Bursa, M., Holzinger, A., Khuri, S. (eds.) ITBAM 2016. LNCS, vol. 9832, pp. 49–66. Springer, Cham (2016). doi:10.1007/978-3-319-43949-5_4
34. Bleiholder, J., Naumann, F.: Data fusion. ACM Comput. Surv. (CSUR) **41**, 1–41 (2008)
35. Lafon, S., Keller, Y., Coifman, R.R.: Data fusion and multicue data matching by diffusion maps. IEEE Trans. Pattern Anal. Mach. Intell. **28**, 1784–1797 (2006)
36. Blanchet, L., Smolinska, A.: Data fusion in metabolomics and proteomics for bio-marker discovery. In: Jung, K. (ed.) Statistical Analysis in Proteomics. MMB, vol. 1362, pp. 209–223. Springer, New York (2016). doi:10.1007/978-1-4939-3106-4_14
37. Hastie, T., Tibshirani, R., Friedman, J.: The Elements of Statistical Learning. Springer, New York (2009)
38. Bishop, C.M.: Pattern Recognition and Machine Learning (2006)
39. Burges, C.J.: A tutorial on support vector machines for pattern recognition. Data Min. Knowl. Disc. **2**, 121–167 (1998)
40. Kotropoulos, C., Pitas, I.: Segmentation of ultrasonic images using support vector machines. Pattern Recogn. Lett. **24**, 715–727 (2003)
41. Ricci, E., Perfetti, R.: Retinal blood vessel segmentation using line operators and support vector classificatio. IEEE Trans. Med. Imaging **26**, 1357–1365 (2007)
42. Orlando, J.I., Blaschko, M.: Learning fully-connected CRFs for blood vessel seg-mentation in retinal images. In: Golland, P., Hata, N., Barillot, C., Hornegger, J., Howe, R. (eds.) MICCAI 2014. LNCS, vol. 8673, pp. 634–641. Springer, Cham (2014). doi:10.1007/978-3-319-10404-1_79
43. Bauer, S., Nolte, L.-P., Reyes, M.: Fully automatic segmentation of brain tumor images using support vector machine classification in combination with hierar-chical conditional random field regularization. In: Fichtinger, G., Martel, A., Peters, T. (eds.) MICCAI 2011. LNCS, vol. 6893, pp. 354–361. Springer, Hei-delberg (2011). doi:10.1007/978-3-642-23626-6_44
44. El-Naqa, I., Yang, Y., Wernick, M.N., Galatsanos, N.P., Nishikawa, R.M.: A sup-port vector machine approach for detection of microcalcifications. IEEE Trans. Med. Imaging **21**, 1552–1563 (2002)
45. Han, J.W., Breckon, T.P., Randell, D.A., Landini, G.: The application of support vector machine classification to detect cell nuclei for automated microscopy. Mach. Vis. Appl. **23**, 15–24 (2012)
46. Breiman, L.: Random forests. Mach. Learn. **45**, 4–32 (2001)
47. Criminisi, A., Jamie, S. (eds.): Decision Forests for Computer Vision and Medical Image Analysis. Springer, London (2013)

48. Zikic, D., Glocker, B., Konukoglu, E., Criminisi, A., Demiralp, C., Shotton, J., Thomas, O.M., Das, T., Jena, R., Price, S.J.: Decision forests for tissue-specific segmentation of high-grade gliomas in multi-channel MR. In: Ayache, N., Delingette, H., Golland, P., Mori, K. (eds.) MICCAI 2012. LNCS, vol. 7512, pp. 369–376. Springer, Heidelberg (2012). doi:10.1007/978-3-642-33454-2_46
49. Glocker, B., Pauly, O., Konukoglu, E., Criminisi, A.: Joint classification-regression forests for spatially structured multi-object segmentation. In: Fitzgibbon, A., Lazebnik, S., Perona, P., Sato, Y., Schmid, C. (eds.) ECCV 2012. LNCS, vol. 7575, pp. 870–881. Springer, Heidelberg (2012). doi:10.1007/978-3-642-33765-9_62
50. Richmond, D., Kainmueller, D., Glocker, B., Rother, C., Myers, G.: Uncertainty-driven forest predictors for vertebra localization and segmentation. In: Navab, N., Hornegger, J., Wells, W.M., Frangi, A.F. (eds.) MICCAI 2015. LNCS, vol. 9349, pp. 653–660. Springer, Cham (2015). doi:10.1007/978-3-319-24553-9_80
51. Criminisi, A.: Anatomy detection and localization in 3D medical images. In: Criminisi, A., Shotton, J. (eds.) Decision Forests for Computer Vision and Medical Image Analysis. Advances in Computer Vision and Pattern Recognition. Springer, London (2013)
52. Štern, D., Ebner, T., Urschler, M.: From local to global random regression forests: exploring anatomical landmark localization. In: Ourselin, S., Joskowicz, L., Sabuncu, M.R., Unal, G., Wells, W. (eds.) MICCAI 2016. LNCS, vol. 9901, pp. 221–229. Springer, Cham (2016). doi:10.1007/978-3-319-46723-8_26
53. Štern, D., Ebner, T., Urschler, M.: Automatic localization of locally similar structures based on the scale-widening random regression forest. In: IEEE International Symposium on Biomedical Imaging (2017)
54. Hebb, D.: The Organization of Behavior. Wiley, New York (1949)
55. McCulloch, W.S., Pitts, W.: A logical calculus of the ideas immanent in nervous activity. Bull. Mathe. Biophys. 5, 115–133 (1943)
56. LeCun, Y., Bottou, L., Bengio, Y., Haffner, P.: Gradient-based learning applied to document recognition. Proc. IEEE 86, 2278–2324 (1998)
57. Krizhevsky, A., Sutskever, I., Hinton, G.E.: Imagenet Classification with Deep Convolutional Neural Networks. In: Advances in Neural Information Processing Systems (2012)
58. Singh, D., Merdivan, E., Psychoula, I., Kropf, J., Hanke, S., Geist, M., Holzinger, A.: Human activity recognition using recurrent neural networks. In: Holzinger, A., Kieseberg, P., Tjoa, A.M., Weippl, E. (eds.) CD-MAKE 2017. LNCS, vol. 10410, pp. 267–274. Springer, Cham (2017). doi:10.1007/978-3-319-66808-6_18
59. Bengio, Y., Courville, A., Vincent, P.: Representation learning: A review and new perspectives. IEEE Trans. Pattern Anal. Mach. Intell. 35, 1798–1828 (2013)
60. Liu, Y., Gadepalli, K., Norouzi, M., Dahl, G.E., Kohlberger, T., Boyko, A., Venugopalan, S., Timofeev, A., Nelson, P.Q., Corrado, G.S., Hipp, J.D., Peng, L., Stumpe, M.C.: Detecting cancer metastases on gigapixel pathology images. arXiv: 1703.02442 (2017)
61. Esteva, A., Kuprel, B., Novoa, R.A., Ko, J., Swetter, S.M., Blau, H.M., Thrun, S.: Dermatologist-level classification of skin cancer with deep neural networks. Nature 542, 115–118 (2017)
62. Ronneberger, O., Fischer, P., Brox, T.: U-Net: convolutional networks for biomedical image segmentation. In: Proceedings Medical Image Computing and Computer-Assisted Intervention (2015)
63. Pan, S.J., Yang, Q.: A survey on transfer learning. IEEE Trans. Knowl. Data Eng. 22, 1345–1359 (2010)

64. Cai, J., Lu, L., Xie, Y., Xing, F., Yang, L.: Pancreas segmentation in MRI using graph-based decision fusion on convolutional neural networks. In: Descoteaux, M., Maier-Hein, L., Franz, A., Jannin, P., Collins, D.L., Duchesne, S. (eds.) MIC-CAI 2017. LNCS, vol. 10435, pp. 674–682. Springer, Cham (2017). doi:10.1007/978-3-319-66179-7_77

65. Payer, C., Štern, D., Bischof, H., Urschler, M.: Regressing heatmaps for multiple landmark localization using CNNs. In: Ourselin, S., Joskowicz, L., Sabuncu, M.R., Unal, G., Wells, W. (eds.) MICCAI 2016. LNCS, vol. 9901, pp. 230–238. Springer, Cham (2016). doi:10.1007/978-3-319-46723-8_27

66. Rozantsev, A., Lepetit, V., Fua, P.: On rendering synthetic images for training an object detector. Comput. Vis. Image Underst. **137**, 24–37 (2015)

67. Goodfellow, I., Pouget-Abadie, J., Mirza, M., Xu, B., Warde-Farley, D., Ozair, S., Courville, A., Bengio, Y.: Generative adversarial nets. In: Advances in Neural Information Processing Systems (2014)

68. Nie, D., Trullo, R., Petitjean, C., Ruan, S., Shen, D.: Medical image synthesis with context-aware generative adversarial networks. arXiv:1612.05362 (2016). Accepted MICCAI'17

69. Schmidhuber, J.: Deep learning in neural networks: An overview. Neural Netw. **61**, 85–117 (2015)

70. Goodfellow, I., Bengio, Y., Courville, A.: Deep Learning. MIT Press (2016)

71. Malle, B., Kieseberg, P., Schrittwieser, S., Holzinger, A.: Privacy aware machine learning and the right to be forgotten. ERCIM News (Special Theme: Machine Learning) **107**, 22–23 (2016)

72. Fosch Villaronga, E., Kieseberg, P., Li, T.: Humans forget, machines remember: Artificial intelligence and the right to be forgotten. Computer Security Law Review (2017)

73. Malle, B., Giuliani, N., Kieseberg, P., Holzinger, A.: The more the merrier - federated learning from local sphere recommendations. In: Holzinger, A., Kieseberg, P., Tjoa, A.M., Weippl, E. (eds.) CD-MAKE 2017. LNCS, vol. 10410, pp. 367–373. Springer, Cham (2017). doi:10.1007/978-3-319-66808-6_24

74. Dwork, C.: Differential privacy: a survey of results. In: Agrawal, M., Du, D., Duan, Z., Li, A. (eds.) TAMC 2008. LNCS, vol. 4978, pp. 1–19. Springer, Heidelberg (2008). doi:10.1007/978-3-540-79228-4_1

75. Kieseberg, P., Hobel, H., Schrittwieser, S., Weippl, E., Holzinger, A.: Protecting anonymity in data-driven biomedical science. In: Holzinger, A., Jurisica, I. (eds.) Interactive Knowledge Discovery and Data Mining in Biomedical Informatics. LNCS, vol. 8401, pp. 301–316. Springer, Heidelberg (2014). doi:10.1007/978-3-662-43968-5_17

76. Schrittwieser, S., Kieseberg, P., Echizen, I., Wohlgemuth, S., Sonehara, N., Weippl, E.: An algorithm for k-anonymity-based fingerprinting. In: Shi, Y.Q., Kim, H.-J., Perez-Gonzalez, F. (eds.) IWDW 2011. LNCS, vol. 7128, pp. 439–452. Springer, Heidelberg (2012). doi:10.1007/978-3-642-32205-1_35

77. Kieseberg, P., Schrittwieser, S., Mulazzani, M., Echizen, I., Weippl, E.: An algorithm for collusion-resistant anonymization and fingerprinting of sensitive microdata. Electron. Markets **24**, 113–124 (2014)

78. Haerder, T., Reuter, A.: Principles of transaction-oriented database recovery. ACM Comput. Surv. (CSUR) **15**, 287–317 (1983)

79. Bayer, R., McCreight, E.: Organization and maintenance of large ordered indexes. In: Broy, M., Denert, E. (eds.) Software Pioneers, pp. 245–262. Springer, Heidelberg (2002)

80. Fruhwirt, P., Kieseberg, P., Weippl, E.: Using internal MySQL/InnoDB B-tree index navigation for data hiding. In: Peterson, G., Shenoi, S. (eds.) DigitalForensics 2015. IAICT, vol. 462, pp. 179–194. Springer, Cham (2015). doi:10.1007/978-3-319-24123-4_11

81. Kieseberg, P., Schrittwieser, S., Mulazzani, M., Huber, M., Weippl, E.: Trees cannot lie: Using data structures for forensics purposes. In: Intelligence and Security Informatics Conference (EISIC), 2011 European, pp. 282–285. IEEE (2011)

82. Pantazos, K., Lauesen, S., Lippert, S.: De-identifying an EHR database-Anonymity, correctness and readability of the medical record. Stud. Health Technol. Inf. **169**, 862–866 (2011)

83. Neamatullah, I., Douglass, M.M., Lehman, L.W.H., Reisner, A., Villarroel, M., Long, W.J., Szolovits, P., Moody, G.B., Mark, R.G., Clifford, G.D.: Automated de-identification of free-text medical records. BMC Med. Inform. Decis. Mak. **8**, 32 (2008)

84. Al-hegami, A.S.: A biomedical named entity recognition using machine learning classifiers and rich feature set. Int. J. Comput. Sci. Netw. Secur. **17**, 170–176 (2017)

85. Settles, B.: Biomedical named entity recognition using conditional random fields and rich feature sets. In: International Joint Workshop on Natural Language Processing in Biomedicine and its Applications, pp. 104–107 (2004)

86. Mavromatis, G.: Biomedical named entity recognition using neural networks **2015**, 1–9 (2015)

87. Goldberg, Y., Levy, O.: word2vec explained: deriving Mikolov et al. negative-sampling word-embedding method. arXiv preprint arXiv:1402.3722 (2014)

88. Sweeney, L.: k-anonymity: A model for protecting privacy. Int. J. Uncertainty, Fuzziness and Knowl.-Based Syst. **10**, 557–570 (2002)

89. Machanavajjhala, A., Gehrke, J., Kifer, D., Venkitasubramaniam, M.: l-diversity: Privacy beyond k-anonymity. In: Proceedings of the 22nd International Conference on Data Engineering, ICDE 2006, p. 24. IEEE (2006)

90. Li, N., Li, T., Venkatasubramanian, S.: t-closeness: Privacy beyond k-anonymity and l-diversity. In: IEEE 23rd International Conference on Data Engineering, ICDE 2007, pp. 106–115. IEEE (2007)

91. Nergiz, M.E., Atzori, M., Clifton, C.: Hiding the presence of individuals from shared databases. In: Proceedings of the 2007 ACM SIGMOD International Conference on Management of Data, pp. 665–676. ACM (2007)

92. Wong, R.C.W., Li, J., Fu, A.W.C., Wang, K.: (α, k)-anonymity: an enhanced k-anonymity model for privacy preserving data publishing. In: Proceedings of the 12th ACM SIGKDD International Conference on Knowledge Discovery and Data Mining, pp. 754–759. ACM (2006)

93. Campan, A., Truta, T.M.: Data and structural k-Anonymity in social networks. In: Bonchi, F., Ferrari, E., Jiang, W., Malin, B. (eds.) PInKDD 2008. LNCS, vol. 5456, pp. 33–54. Springer, Heidelberg (2009). doi:10.1007/978-3-642-01718-6_4

94. Malle, B., Kieseberg, P., Weippl, E., Holzinger, A.: The right to be forgotten: towards machine learning on perturbed knowledge bases. In: Buccafurri, F., Holzinger, A., Kieseberg, P., Tjoa, A.M., Weippl, E. (eds.) CD-ARES 2016. LNCS, vol. 9817, pp. 251–266. Springer, Cham (2016). doi:10.1007/978-3-319-45507-5_17

95. Malle, B., Kieseberg, P., Holzinger, A.: DO NOT DISTURB? classifier behavior on perturbed datasets. In: Holzinger, A., Kieseberg, P., Tjoa, A.M., Weippl, E. (eds.) CD-MAKE 2017. LNCS, vol. 10410, pp. 155–173. Springer, Cham (2017). doi:10.1007/978-3-319-66808-6_11

96. Rafique, A., Azam, S., Jeon, M., Lee, S.: Face-deidentification in images using restricted boltzmann machines. In: ICITST, pp. 69–73 (2016)

97. Chi, H., Hu, Y.H.: Face de-identification using facial identity preserving features. In: 2015 IEEE Global Conference on Signal and Information Processing, Global-SIP 2015, pp. 586–590 (2016)

98. Yu, F., Fienberg, S.E., Slavković, A.B., Uhler, C.: Scalable privacy-preserving data sharing methodology for genome-wide association studies. J. Biomed. Inform. 50, 133–141 (2014)

99. Simmons, S., Sahinalp, C., Berger, B.: Enabling privacy-preserving GWASs in heterogeneous human populations. Cell Syst. 3, 54–61 (2016)

100. Im, H.K., Gamazon, E.R., Nicolae, D.L., Cox, N.J.: On sharing quantitative trait GWAS results in an era of multiple-omics data and the limits of genomic privacy. Am. J. Hum. Genet. 90, 591–598 (2012)

101. Knoppers, B.M., Dove, E.S., Litton, J.E., Nietfeld, J.J.: Questioning the limits of genomic privacy. Am. J. Hum. Genet. 91, 577–578 (2012)

102. Aggarwal, C.C., Li, Y., Philip, S.Y.: On the hardness of graph anonymization. In: 2011 IEEE 11th International Conference on Data Mining (ICDM), pp. 1002–1007. IEEE (2011)

103. Baxter, J.: A model of inductive bias learning. J. Artif. Intell. Res. 12, 149–198 (2000)

104. Evgeniou, T., Pontil, M.: Regularized multi-task learning. In: Proceedings of the Tenth ACM SIGKDD International Conference on Knowledge Discovery and Data Mining, pp. 109–117. ACM (2004)

105. Weinberger, K.Q., Saul, L.K.: Distance metric learning for large margin nearest neighbor classification. J. Mach. Learn. Res. 10, 207–244 (2009)

106. Parameswaran, S., Weinberger, K.Q.: Large margin multi-task metric learning. In: Lafferty, J., Williams, C., Shawe-Taylor, J., Zemel, R., Culotta, A. (eds.) Advances in Neural Information Processing Systems 23 (NIPS 2010), pp. 1867–1875 (2010)

107. McCloskey, M., Cohen, N.J.: Catastrophic interference in connectionist networks: The sequential learning problem. In: Bower, G.H. (ed.) The Psychology of Learning and Motivation, vol. 24, pp. 109–164. Academic Press, San Diego (1989)

108. French, R.M.: Catastrophic forgetting in connectionist networks. Trends Cogn. Sci. 3, 128–135 (1999)

109. Goodfellow, I.J., Mirza, M., Xiao, D., Courville, A., Bengio, Y.: An empirical investigation of catastrophic forgeting in gradient-based neural networks. arXiv:1312.6211v3 (2015)

110. Taylor, M.E., Stone, P.: Transfer learning for reinforcement learning domains: A survey. J. Mach. Learn. Res. 10, 1633–1685 (2009)

111. Sycara, K.P.: Multiagent systems. AI Mag. 19, 79 (1998)

112. Lynch, N.A.: Distributed Algorithms. Morgan Kaufmann, San Francisco (1996)

113. DeGroot, M.H.: Reaching a consensus. J. Am. Stat. Assoc. 69, 118–121 (1974)

114. Benediktsson, J.A., Swain, P.H.: Consensus theoretic classification methods. IEEE Trans. Syst. Man Cybern. 22, 688–704 (1992)

115. Weller, S.C., Mann, N.C.: Assessing rater performance without a gold standard using consensus theory. Med. Decis. Making 17, 71–79 (1997)

116. Olfati-Saber, R., Fax, J.A., Murray, R.M.: Consensus and cooperation in networked multi-agent systems. Proc. IEEE 95, 215–233 (2007)

117. Roche, B., Guegan, J.F., Bousquet, F.: Multi-agent systems in epidemiology: a first step for computational biology in the study of vector-borne disease transmission. BMC Bioinf. **9** (2008)
118. Kok, J.R., Vlassis, N.: Collaborative multiagent reinforcement learning by payoff propagation. J. Mach. Learn. Res. **7**, 1789–1828 (2006)

Comparison of Public-Domain Software and Services For Probabilistic Record Linkage and Address Standardization

Sou-Cheng T. Choi[1,2]([✉]), Yongheng Lin[3], and Edward Mulrow[3]

[1] Illinois Institute of Technology, Chicago, IL, USA
schoi32@iit.edu
[2] Allstate Corporation, Chicago, IL, USA
[3] NORC at the University of Chicago, Chicago, IL, USA

Abstract. Probabilistic record linkage (PRL) refers to the process of matching records from various data sources such as database tables with some missing or corrupted index values. Human is often involved in a loop to review cases that an algorithm cannot match. PRL can be applied to join or de-duplicate records, or to impute missing data, resulting in better overall data quality. An important subproblem in PRL is to parse a field such as address into its components, e.g., street number, street name, city, state, and zip code. Various data analysis techniques such as natural language processing and machine learning methods are often gainfully employed in both PRL and address standardization to achieve higher accuracies of linking or prediction. This work compares the performance of four reputable PRL packages freely available in the public domain, namely FRIL, Link Plus, R RecordLinkage, and SERF. In addition, we evaluate the baseline performance and sensitivity of four address-parsing web services including the Data Science Toolkit, Geocoder.us, Google Maps APIs, and the U.S. address parser. Finally, we present some of the strengths and limitations of the software and services we have evaluated.

Keywords: Probabilistic record linkage · Heterogeneous data · Address standardization · Fellegi-sunter model · Geocoding

1 Introduction

Record linkage (RL) is defined as linking, matching, or deduplication of data records from one or multiple data sources in the presence of some missing data in index key(s). Two comprehensive works on the subject are Christen (2012) [1] and Herzog et al. (2007) [2]. RL is closely related to *entity resolution* (ER), which is also known as deduplication of databases records [3]. ER, in its most general sense, is "the process of determining whether two references to real-world objects are referring to the same object or to different objects" [4, Chapter 1]. As the underlying data sources in RL we consider could be flat files, structured tables from databases or data warehouses, or unstructured data from social media

© Springer International Publishing AG 2017
A. Holzinger et al. (Eds.): Integrative Machine Learning, LNAI 10344, pp. 51–66, 2017.
https://doi.org/10.1007/978-3-319-69775-8_3

or the internet, we do not need to distinguish RL from ER unless otherwise specified. It is also clear that RL or ER often appear as an integral part of big-data management.

RL emerged in public health and policy studies as early as year 1946 [5,6]. Today precision medicine is one of the grand challenges in medical sciences and public health; it refers to the medical science and preventive measures that are personalized for an individual considering one's characteristics such as genes, environment, and lifestyle. Precision medicine is largely data-driven; RL and particularly, probabilistic record linkage (PRL) [7], can be readily applied for linking patient records from various data sources such as cancer and mortality registries, pharmacy records, and digitalized medical transcriptions [8,9].

There are many mission-critical scientific and business applications of PRL. In survey research, PRL is often one of the data preprocessing steps—along with coding, imputation, and weighting—for enhancing data quality so that later statistical analysis can be more accurate [10]. For instance, the Census Bureau's Person Identification Validation System is a production system that uses PRL [11,12] for matching person records arising from censuses records, survey datasets, administrative records, or commercial files.

There are two main kinds of matching in RL: exact and statistical [2]. Exact matching is best known for PRL, mostly associated with the celebrated Fellegi-Sunter model [7]. Even though many novel machine-learning methods have been developed for statistical matching, the Fellegi-Sunter method remains essential, employed by many government agencies [1].

In this study, we evaluate four PRL software, namely, Fine-grained Record Integration and Linkage (FRIL) [13], Link Plus [14], R RecordLinkage [15], and the Stanford Entity Resolution Framework (SERF) [3]. We choose these software based on a few criteria: available in public domains, easy to install and use, free of charge, and preferably open source. We do not consider commercial software such as AutoMatch [16] and DataMatch [17]. Likewise, we do not include Link King [18] because it requires a SAS license [19]. We approach the software as most first-time users without expert knowledge about the software probably would, using the products without excessive fine tuning input parameters for optimal performance. For comparison purpose, we used the dataset `RLdata10000` [15].

A frequently encountered sub-problem of address-based PRL is parsing or standardization of addresses. Hence we have selected multiple packages with Python interfaces: U.S. address parser [20], Google Maps Geocoding APIs [21], Geocoder.us [22], and Data Science Toolkit [23]. For testing, we employed a test dataset of 20 000 addresses without personally identifiable information (PII); this dataset is sampled from an USPS address database [24], which contains millions of relatively clean business and residential addresses in America. To gain a sense of the robustness of the software services, we introduced noises to the address fields and repeated the standardization procedures.

Here is an outline of this paper: In Sect. 2, we briefly recapitulate the mathematical models that underlie PRL. In Sect. 3, we report on the empirical performance of the chosen PRL software. In Sect. 4, we summarize the results of our testing on the selected address standardization services. We conclude in Sect. 5 with a summary of our learning and thoughts. We also make a couple of recommendations for future research work.

2 Basic Mathematical Models

Let a and b be two records from sets $\mathbf{A}$ and $\mathbf{B}$ respectively. Define two disjoint sets $\mathbf{M}$ and $\mathbf{U}$ such that $\mathbf{A} \times \mathbf{B} = \mathbf{M} \bigcup \mathbf{U}$, i.e.,

$$\mathbf{A} \times \mathbf{B} = \{(a,b) : a \in \mathbf{A}, b \in \mathbf{B}\},$$
$$\mathbf{M} = \{(a,b) \in \mathbf{A} \times \mathbf{B} : a = b\} = \langle\text{matched set}\rangle,$$
$$\mathbf{U} = \{(a,b) \in \mathbf{A} \times \mathbf{B} : a \neq b\} = \langle\text{unmatched set}\rangle.$$

Let γ be a comparison N-vector of given a and b, where N is a positive integer, usually small. Let $\mathbf{T} = \{\gamma : (a,b) \in A \times B\}$. If elements of γ are binary values, then $\mathbf{T}$ contains at most 2^N unique elements.

Here is a simple example for illustration. Suppose a has five fields (Anne, Ford, IL, 1980, \$70,000) and b has five fields (Ann, Ford, IL, 1980, married). Let $N = 4$ and the comparison vector of a and b to be based on the first N fields of the records. The simplest way to define γ is to use *exact matching*: $\gamma = (\gamma_i)_{i=1}^N$ where γ_i is 1 if $a_i = b_i$, or 0 otherwise. In this case, $\gamma = (0,1,1,1)$. In practice, exact matching is often found to be too restrictive as it would not tolerate any typographical errors in names or small differences in values.

More generally, we can define γ_i to be 1 if $d(a_i, b_i)$, some distance measure between a_i and b_i, is less than some given threshold τ_i, or define γ_i to be 0 otherwise. For example, suppose $\tau_1 := 3$ and with edit distance $d(a_1, b_1) = d(\text{Anne}, \text{Ann}) = 1 < \tau_1$, we have $\gamma_1 = 1$ and $\gamma = (1,1,1,1)$. In this case, we can consider a and b matching and merge a and b to form a more informative record, (Anne, Ford, IL, 1980, \$70,000, married).

2.1 SERF

SERF requires *normalized* distance functions and user-input threshold values τ_i between 0 and 1 on each matching field. It uses the decision rule that if $d(a_i, b_i) \leq \tau_i$ for all $i = 1, \ldots, N$, then consider $(a,b) \in \mathbf{M}$, i.e., a and b matching (or nonmatching otherwise). When two records are matched, the iterative algorithm merges them immediately before comparing next record pair. It assumes only three data types for all fields and the distance functions are defined accordingly:

1. If r and s are non-negative real numbers and they are not both zero, then $d(r,s) = \frac{|r-s|}{\max(r,s)}$.
2. If r and s are values of a categorical variable, then use exact matching.
3. If r and s are text strings, then $d(r,s)$ is defined to be the Jaro-Winkler distance [25] of r and s.

2.2   The Fellegi-Sunter Model

The Fellegi-Sunter model decides for every distinct record pair $(a, b) \in A \times B$ exactly one of the following three cases:

$A_1 = \langle$model deciding (a, b) are a match$\rangle$,

$A_2 = \langle$model deciding(a, b) are a potential match, to be manually reviewed$\rangle$,

$A_3 = \langle$model deciding(a, b) are not a match$\rangle$.

Define m- and u-probabilities of γ associated with (a, b) as:

$$m(\gamma) = P(\gamma|(a, b) \in M),$$
$$u(\gamma) = P(\gamma|(a, b) \in U).$$

Assuming conditional independence of elements in γ, the model defines S, a matching score for (a, b) as the logarithm of the ratio of the m- and u-probabilities:

$$S \equiv \log_2 \left(\frac{m(\gamma)}{u(\gamma)} \right) = \sum_{i=1}^{N} \log_2 \frac{P(\gamma_i|(a, b) \in M)}{P(\gamma_i|(a, b) \in U)} =: \sum_{i=1}^{N} w_i.$$

If $S \in \mathbb{R}$ is large positive, then (a, b) are more likely to be matching. On the other hand, if S is large negative, then (a, b) are more likely to be nonmatching. Suppose we are given lower and upper bounds L and U such that $L < U$. The Fellegi-Sunter model decides that (a, b) is in A_1 if $S \geq U$, in A_2 if $S \in (L, U)$, or in A_3 if $S \leq L$. The bounds L and U are determined by a priori probabilities of errors, μ and $\lambda \in [0, 1]$, which are also known as false positive rate (FPR) and false negative rate (FNR), respectively:

$$\mu = \sum_{\gamma \in \mathbf{T}} u(\gamma) P(A_1|\gamma) = P(A_1|U),$$
$$\lambda = \sum_{\gamma \in \mathbf{T}} m(\gamma) P(A_3|\gamma) = P(A_3|M).$$

Denote a linkage rule as $L(\mu, \lambda, \mathbf{T})$. The optimal linkage rule is defined as one that minimizes the number of cases in A_2, which often have to be reviewed by human efforts:

$$L^* = \arg\min_L P(A_2|L).$$

For evaluation of the Fellegi-Sunter model, if datasets with labels of matching or not for each record pairs are available, we can use a special form of confusion matrix (often used in supervised machine learning) as shown in Table 1, where TP, FP, FN, and TN are counts of true positive, false positive, false negative, and true negative, respectively; $A_{2,1}$ and $A_{2,2}$ sum to the number of records in A_2. Human is often involved to review the cases that fall in A_2 to manually reclassify them to A_1 or A_3.

Table 1. Confusion matrix for evaluating the practical performance of the Fellegi-Sunter model used for PRL on dataset records with labels of matching or not.

	A_1	A_2	A_3
$(a,b) \in M$	TP	$A_{2,1}$	FN
$(a,b) \in U$	FP	$A_{2,2}$	TN

3 PRL Software

We evaluate R RecordLinkage (version 0.4-8), FRIL (version 2.1.5), Link Plus (version 2.0), and SERF (version 0.1) using the latest versions available in the public domain as of June 2016. We ran the programs on a common dataset RLdata10000 provided by the R RecordLinkage package. The dataset contains a total of 10,000 records and seven fields. The first field is a running identification numbers for the records; the next four fields are European name parts; and the last three are birthday information. Here are the first three records in the dataset (note that NA means that data is not available):

id	fname_c1	fname_c2	lname_c1	lname_c2	by	bm	bd
1	FRANK	NA	MUELLER	NA	1967	9	27
2	MARTIN	NA	SCHWARZ	NA	1967	2	17
3	HERBERT	NA	ZIMMERMANN	NA	1961	11	6

A total of one thousand pairs of records are *known* to be duplicates in this dataset. The following are two example pairs of matching records with small differences in first name and birth year, respectively:

id	fname_c1	fname_c2	lname_c1	lname_c2	by	bm	bd
4	HANS	NA	SCHMITT	NA	1945	8	14
1957	HRANS	NA	SCHMITT	NA	1945	8	14
12	STEFAN	NA	RICHTER	NA	1943	3	22
4269	STEFAN	NA	RICHTER	NA	1946	3	22

A total of $n(n-1)/2 = 49995000$ (for $n = 10000$) pairwise comparisons are made. In Table 2, we summarize the performance of the aforementioned products. We can see that R RecordLinkage and FRIL attained the highest true match counts and F_1 scores, with FRIL being the fastest. R RecordLinkage, being the provider of the dataset, might have optimized itself against the European-centric dataset. The dataset is also considered small in real-world problems. FRIL and SERF, which are Java based, will probably scale substantially better than R RecordLinkage on larger datasets.

In the rest of this section, we detail some features of each software, as well as our test process yielding the results in Table 2.

Table 2. Performance of selected record linkage software. The number of cases that fall in A_2 is zero in all cases. F_1 score, a measure of accuracy, is defined as $(2TP)/(2TP + FP + FN)$ and run time does not include time for computing the confusion matrix.

Software	TP	TN	FP	FN	$|A_2|$	F_1 score	Run time (seconds)
FRIL	959	49993946	54	41	0	95.3 %	8.5
Link plus	586	49993688	312	414	0	61.7 %	19.0
R RecordLinkage	970	49993931	69	30	0	95.1 %	65.8
SERF	749	49993833	167	251	0	78.2%	46.8

3.1 FRIL

FRIL [13] is an open-source, well-documented [26] package with both binary and source distribution. It is Java based and can be installed on Windows, Linux/Unix, or Mac platforms. It comes with a simple graphical user interface (GUI). A record linkage or deduplication workflow can be captured in an XML configuration file for reproducibility. For deduplication of a dataset, a user need to specify a weight, w_i, for each of the N matching attributes and a value for an input parameter duplicate acceptance level, denoted by δ here. Two records are considered a duplicate of each other if $\sum_{i=1}^{N} w_i d(a_i, b_i) > \delta$ [26, p. 38–41].

We executed FRIL on RLdata10000 without blocking (by using a dummy attribute with the same value for every record) and set the parameter δ to be 80. In addition, we set the weight of each name related attribute to be 16 and the weight of each birthday related attribute to be 12, summing to 100 in total. We used the default distance function for each attribute, Equal fields boolean distance, which is the same as exact matching. The software automatically took advantage of all the CPU cores available in the deduplication process; in our case, eight (logical) cores were used. We note that FRIL identified six triplets of matching records, among other duplicated pairs. One of the triplets is shown below; the first and the third records are known to be duplicates, but the second record is not a replica of any other records in the dataset:

id	fname_c1	fname_c2	lname_c1	lname_c2	by	bm	bd
317	THOMAS	NA	SCHMID	NA	1993	3	5
2156	THOMAS	NA	BECKER	NA	1993	3	5
4940	THOMAVS	NA	SCHMID	NA	1993	3	5

3.2 Link Plus

Link Plus [14] is regrettably *not* open source and installable on only Windows machines. However, the application has a user friendly GUI and each screen has an easily accessible help menu. Moreover, scientific workflows can be saved in a configuration .cfg file. A deduplication process requires a user-specified value for the input parameter, Cutoff Value.

For the dataset `RLdata10000`, we chose `Cutoff Value` to be 10. We used the matching methods of *first name* and *last name* for the name fields, and *generic string* for the birthday related fields. We did not use blocking. A total of 898 pairs were matched up by Link Plus. A duplicate may be matched up with multiple records; for example, the record whose `id` = 2219 is considered by the algorithm a close copy of three records whose `id` are 22, 1295, and 3008:

id	fname_c1	fname_c2	lname_c1	lname_c2	by	bm	bd
2219	GUENTHER	NA	MUELLER	NA	1955	11	1
22	GUENTHER	NA	MUELLER	NA	1994	11	17
1295	GUENTHER	NA	MUELLER	NA	1993	11	19
3008	GUENTHER	NA	MUELLER	NA	1985	11	16

3.3   R RecordLinkage

R RecordLinkage is a package written in R. It is open source accompanied by comprehensive documentation. For deduplication, it requires user input for a parameter called `threshold.upper`. There are a number of useful features that come with the package: blocking, Soundex algorithm, Levenshtein distance, and Jaro-Winkler algorithms (implemented in C). The package also has a stochastic framework for computing weights with an EM algorithm (implemented in C) and provides easy access to several modern machine learning methods.

The following is our R code for deduplicating the records in `RLdata10000`, with `threshold.upper` set to be 0.53:

```
library(RecordLinkage)
data(RLdata10000)
system.time({ rpairs = compare.dedup(RLdata10000,
            identity = identity.RLdata10000) })
system.time({ rpairs = epiWeights(rpairs) })
res = epiClassify(rpairs, 0.53)
summary(res)
```

3.4   SERF

SERF is an open-source Java package with complete JavaDocs documentation. It requires some, but not too much, configuration and compilation of a JAR (Java archive) file from the source code of SERF; another Java package called Second-String [27]; and a user-written Java class, which is customized for input dataset attributes, extends and implements the SERF Java classes `BasicMatcherMerger` and `MatcherMerger`, respectively. The input files to SERF are required to be transformed into a certain XML format. SERF also need user-input threshold values between zero and one on each of the user-selected fields specified in a configuration file. It matches pairwise records by a few distance functions on the selected fields (e.g., Jaro-Winkler distance function on two strings) whenever all of the measures are smaller than the corresponding thresholds.

We transformed our test dataset, `RLdata10000.rda`, a binary format in R, into csv (for ease of manual inspection) and SERF-required XML formats. We ran SERF on the XML file with minimally tuned threshold values. The resultant outputs are again persisted in XML file format, with matched records "nested" together. We developed some Python tools for transforming data between CSV and SERF XML formats; such transformations took an insignificant amount of execution time for the test set. In a configuration file (`example12.conf`), we specify the paths of input and output files, name of our Java class customized for attributes of `RLdata10000`, and set a threshold value for each attribute as follows:

```
FileSource=~/norc_prl/Software/norc-serf/example/RLdata10000b.xml
OutputFile=~/norc_prl/Software/norc-serf/example/output12.xml
MatcherMerger=serf.data.RLdataMatcherMerger
YearThreshold = .999
MonthThreshold = .7
DayThreshold = .3
LnameThreshold = .94
FnameThreshold = .9
Lname2Threshold = .33
Fname2Threshold = .33
```

Lastly, we ran the following command to run SERF for deduplicating the dataset:

```
time java -cp"../libs/norc-serf.jar" serf.ER example12.conf
```

4   Street Standardizers

Address parsing, or address standardization, refers to the process of separating a line or a chunk of text into its components such as street number, street name, city, state, zip code, country, so on and so forth. It is an important process in problems such as record linkage or geocoding (mapping an address to latitudinal and longitudinal coordinates). Often, various modern data analysis techniques such as natural language processing and machine learning methods are gainfully employed to achieve higher accuracies.

For our evaluation of address standardization software, we first performed small tests consisting of a few full and partial addresses, as well as addresses with redundant information. This process enabled us to quickly gain a sense of how well the service providers perform. Only if the services returned satisfactory parsing results for the simple test cases, we considered them for a second-stage large-scale batch tests. Two interesting web services that were not selected into the final tests were Gisgraphy [28] and Yahoo!Placefinder [29]. While we did not install Gisgraphy, its online web interface [30] provides a comparison of its service with other similar providers. With our simple test cases, Gisgraphy consistently returned parsing results less satisfactory than those from, for example, Google

Maps, and Yahoo!Placefinder. During our period of study, Yahoo!Placefinder suspended the service of issuing new application keys for research developers. Hence we also did not explore the service further.

For larger batch tests, we use a fraction of U.S. addresses from the Valassis Database [24]. The advantages of this data source are that it is large, clean, carefully tabulated into standard fields, and has no PII. The address database is partitioned into 23 compressed files (in zip format), each with address records sorted in increasing zip code. We randomly sampled from the first file, resulting in a test set of 20 000 addresses randomly distributed in nine states in the northeastern part of America; see Figs. 3 and 4 for their spatial distribution. For every address, we concatenate values, if they exist, from eleven fields in order: street_num, street_pre_dir, street_name, street_suffix, street_post_dir, unit_type, unit_num, city, state_abbrev, zip, and zip4. For example, 146 SPROUL RD MALVERN PA 19355 1954 is one of the test addresses we used.

For more realistic testing, we randomly selected 10% of all the records in the clean sample and introduced to each selected record up to six random mutations of text characters such as omission, transposition, and repetitions. An example of a mutated address is 21 HOPE FARMS DR FEEDING HILLS MA010300 2[S013 (Cf. the clean address: 21 HOPE FARMS DR FEEDING HILLS MA 01030 2013).

The clean dataset is used to establish baseline performance of a few lightweight address-parsing web services including the U.S. address parser [20], Google Maps geocoding APIs [21], Geocoder.us [22], and Data Science Toolkit [23]. Table 3 above shows that U.S. address and Google Maps achieved the highest baseline accuracies of 99.3% and 80.6%, respectively. They also offered the most robust services on the noisy dataset with accuracies of recovering the original addresses at approximately 95.3% and 80.6%, respectively. In this study, the U.S. address parser is the most accurate and the fastest, whereas the Google Maps is the least sensitive to contaminated data, indicating that its underlying analytics platform may have reliable error correction functions.

Table 3. Performance of various address-parsing web services. Here the accuracy is computed as $\mathrm{TM}/\mathrm{TM} + \mathrm{FM}$, where TM and FM are true and false match counts. The values were refreshed on a day in July of 2017.

Software service	Number of parsed records	True match count	False match count	Accuracy	Time user + sys (seconds)
	Clean; Noisy	Clean; Noisy	Clean; Noisy	Clean; Noisy	Clean; Noisy
U.S. Address	20000; 20000	150 353; 144 215	988; 7125	99.3%; 95.3%	17; 17
Google Maps	20 000; 19 991	122 046; 119 748	29 294; 31 520	80.6%; 79.2%	96; 95
Geocoder.us	20 000; 19 998	80 458; 74 267	50 882; 57 061	61.3%; 56.6%	54; 55
Data Science Toolkit	19 997; 19 480	88 652; 83 500	42 670; 44 473	67.5%; 65.2%	68; 68

For a more thorough sensitivity analysis of the software packages to noise levels, we vary the percentage of noisy records from 10% to 40% in increments of 10%. Figures 1 and 2 summarize the accuracies and CPU execution time of the packages. It is clear that the noisier the data, the less accurate the parsed results are. Neverthess, U.S. address remains the most performant; its worst result is still better than that of the other packages. Run time is however not sensitive to the noise levels.

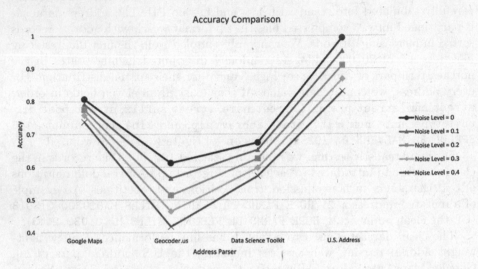

Fig. 1. Comparison of parsing accuracies of our four address standardization packages.

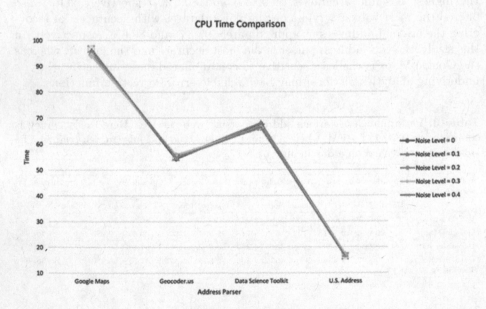

Fig. 2. Comparison of run time (CPU time) of our selected address standardization packages.

We present below features, as well as strengths and limitations of the services we have evaluated.

4.1   U.S Address Parser

DataMade offers a U.S. address parsing APIs service through a free and user friendly Python library called "usaddress." Underlying the library is a machine learning method known as the conditional random fields (CRFs) [31], which are based on probabilistic graphical models [32], often found to be employed in problems of image recognition. For address parsing, the model is capable of learning and prediction of both the order and characteristics of address fields.

4.2   Google Maps APIs

There are multiple Python client packages that connect to the Google Maps API web services. We explored three such services: googlemaps [33], pygeocoder [34], which is built on googlemaps, and lastly, issuing REST APIs to Google Maps [21], e.g., http://maps.googleapis.com/maps/api/geocode/json?address=60605+Illinois+USA. The first two Python APIs require a Google API key as an input, which can be generated online with Google; they performed very similarly. Each of the three services allows up to 2500 free-of-charge, non-commercial HTTP requests per day. They all return parsed results in JSON (JavaScript Object Notation) format. The services are generally reliable and the results are of high quality. Nonetheless, the last four digits of a nine-digit zip code are almost always omitted in the parsed results.

4.3   Geocoder.us

Goecoder.us is a Perl library built upon the U.S. Census Bureau's TIGER/Line (Topologically Integrated Geographic Encoding and Referencing/Line) Files. We simply use Geocoder.us's REST interfaces for address parsing, without installing the whole package and the Census data. Nonetheless, a significant number of addresses were not found. The last four digits of a nine-digit zip code, unit type, and unit numbers are almost always omitted in the parsed results.

4.4   Data Science Toolkit

Data Science Toolkit offers RESTful service similar to that of Google Maps. For instance, the query http://www.datasciencetoolkit.org/maps/api/geocode/json?address=60605+Illinois+USA returns a JSON output of geographical properties and address components. However, the outputs too often drop all zip code, unit number, and unit type.

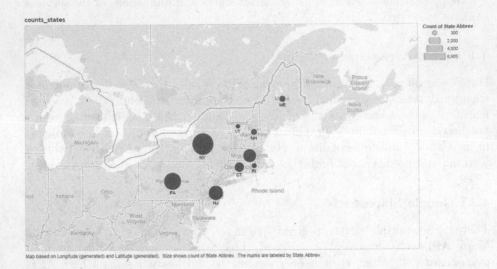

Fig. 3. Spatial distribution of the states of our test sample of 20 000 addresses in U.S. The addresses are associated with only nine states in the northeastern part of the country, namely Connecticut (CT), Massachusetts (MA), Maine (ME), New Hampshire (NH), New Jersey (NJ), New York (NY), Pennsylvania (PA), Rhode Island (RI), and Vermont (VT). The size of a blue circle is proportional to the number of addresses associated with the state it represents. (This figure is produced by Tableau.)

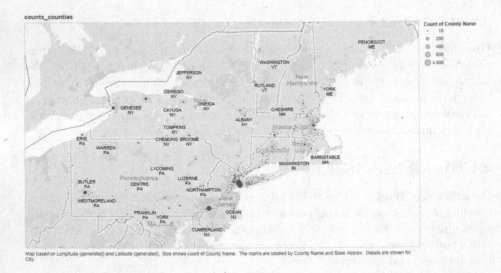

Fig. 4. Spatial distribution of the counties of our 20 000 U.S. address sample.

5   Conclusions

We hope that this article may serve as a brief overview to the two methodologies, namely probabilistic record linkage and address standardization as shown in Sects. 1 and 2. Sections 3 and 4 as well as the code in the accompanied online GitHub repository [35] may serve as quick-start tutorials on the basics of the eight selected related software packages. Alternatively, these materials may be freely used as a basis for further research by us or readers—the last part of this section outlines some potential fruitful research projects of different scope. For comprehensive documentation of the complex software or fine-tuning the (hyper)parameters of the models and applications, users are however urged to refer to the original software manuals or related publications, some of which are listed in the References section.

We have demonstrated the use of freely available software for PRL and online machine-learning services for address standardization. The use of third-party pre-trained learning models built from big data has started mainly with image recognition using deep learning neural network, e.g., Inception [36] trained on the ImageNet [37] with over 14 million images and available from Tensorflow [38]; VGG-Face [39] from MatConvNet [40]. Some of the pre-built models are available as online services or for download. Hence we have been inspired to borrow such practices to standardize unstructured text fields such as addresses. One may reasonably have questions or doubts about the performance or reliability of such services for research or business purposes. One of the goals of this paper is to explore and provide evidence that such online services could be usable but with accuracy measures ranging from poor to excellent as shown in Fig. 1.

As shown, the online services we have studied often work well, returning good-quality data and computational results. However, there are times when the remote services are not available or do not have sufficient capacity. For instance, `geocoder.us` went off line for many days during November and December 2015. A web-service API (application programming interfaces) may not work reliably immediately after the underlying data format or the signature of the API evolves and changes. Clients certainly require stable and high bandwidth internet services in order to effectively interact with the remote servers. Some application packages are non-trivial to install to begin with as they may need databases and compilation of source code, and setting up an environment with complex dependencies. Also, in practice, record linkage usually works with PII. Hence, in addition to speed and accuracy, the choice of software would also depend on whether there is an option for an organization to install the software locally in a private secure environment.

In the rest of this section, we outline a number of potentially fruitful research tasks for future exploration:

1. Exploring more software or services: MapQuest [41], Yahoo!Placefinder [29], and Bing Maps REST Services [42] are some of the reputable geocoding services that we have not explored in depth due to limited resources or unavailability of services during our study period.

2. Cross validation: For our work, we mostly used default values for various matching parameters. In the case of RecordLinkage and SERF, we had quickly tuned the software by trying a few different threshold values. In reality, parameter-tuning via cross validation is often among the most time-consuming but critical tasks of PRL.

3. Phonetics: Soundex and NYSII are two well-established schemes for encoding and comparing common English names close in pronunciation. However, they are challenged by minority or international names that involve non-English alphabets. Some research has indicated that double metaphone [43] is more applicable for linking such records [1]. Some preliminary studies of double metaphone are reported in [44].

4. Linking with social media data: As social media (e.g., Twitter, Facebook) proliferates, linking administrative records with social media data may be fruitful. A use case may be to infer missing values such as gender or age in traditional relational database tables from semi-structured data, often captured in JSON or XML formats, originated from public Facebook or Twitter accounts.

5. Name parsing: Similar to address parsing, name parsing is often found to be another subproblem in probabilistic record linkage. Note that the provider of U.S. address parser also offers Western name parsing service through the Python package "probablepeople" [45].

6. Ensemble learning: To improve the prediction results of pretrained models for address standardization, we could apply ensemble learning such as adaptive boost, so that the combined prediction results could be stronger than any individual method.

7. Software development: we could develop our own models for the problems of interest. In fact, our initial attempts of using LSTM (long short-term memory) neural networks with a CRF layer for entity recognition have shown promising address standardization accuracies surpassing 99% on clean data.

8. Bigger datasets: It would be of interest to many business organizations to evaluate the frameworks on bigger datasets, for instance, wider coverage of U.S. addresses or international addresses beyond the United States.

Acknowledgements. We thank Kirk Wolter, Ned English, and Ilana Ventura for discussion. We also thank Katie Dekker for her expertise in sampling the USPS address database [24]. We are grateful to the feedback from Andreas Holzinger. The first author would like to thank the following people for interesting and inspiring discussion: Forest Gregg, Lulu Kang, Aleksandr Likhterman, Lek-Heng Lim, Dean Resnick, and students enrolled in the research course SCI 498 / MATH 491 Computational Social Sciences, Illinois Institute of Technology, Summer 2016 — in particular, Fabrício Soares deserves special thanks for figuring out the IP address of `geocoder.us` server [22]. Last but not least, we appreciate the assistance from Jack Huang, University of Chicago, in verifying and enhancing our code for address standardization in Summer 2017.

References

1. Christen, P.: Data Matching: Concepts and Techniques for Record Linkage, Entity Resolution, and Duplicate Detection. Springer, Heidelberg (2012)
2. Herzog, T.N., Scheuren, F.J., Winkler, W.E.: Data Quality and Record Linkage Techniques. Springer, New York (2007)
3. Benjelloun, O., Garcia-Molina, H., Kawai, H., Larson, T.E., Menestrina, D., Su, Q., Thavisomboon, S., Widom, J.: Generic entity resolution in the SERF project. IEEE Data Eng. Bull. **29**(2), 13–20 (2006)
4. Talburt, J.R.: Entity Resolution and Information Quality. Elsevier, New York (2011)
5. Dunn, H.L.: Record linkage. Am. J. Public Health Nations Health **36**, 1412–1416 (1946)
6. Schwartz, E.E.: Some observations on the Canadian family allowances program. Soc. Serv. Rev. **20**(4), 451–473 (1946)
7. Fellegi, I.P., Sunter, A.B.: A theory for record linkage. J. Am. Stat. Assoc. **64**, 1183–1210 (1969)
8. Holzinger, A., Jurisica, I.: Interactive Knowledge Discovery and Data Mining in Biomedical Informatics: State-of-the-Art and Future Challenges, vol. 8401. Springer, Heidelberg (2014)
9. Hudson, K., Lifton, R., Patrick-Lake, B.: The precision medicine initiative cohort program – building a research foundation for 21st century medicine (2015). http://acd.od.nih.gov/reports/DRAFT-PMI-WG-Report-9-11-2015-508.pdf
10. Biemer, P.: Introduction to Part 2: Survey processing. In: Pfeffermann, D., Rao, C.R. (eds.) Handbook of Statistics 29A: Sample Surveys: Design, Methods and Applications, pp. 157–162. Elsevier (2009)
11. Wagner, D., Layne, M.: The Person Identification Validation System (PVS): Applying the Center for Administrative Records Research and Applications' (CARRA) record linkage software. CARRA working paper series (2014)
12. Mulrow, E., Mushtaq, A., Pramanik, S., Fontes, A.: Assessment of the US Census Bureau's Person Identification Validation System. Technical report, NORC at the University of Chicago (2011)
13. Jurczyk, P., Lu, J.J., Xiong, L., Cragan, J.D., Correa, A.: FRIL: A tool for comparative record linkage. In: AMIA Annual Symposium Proceedings. American Medical Informatics Association, vol. 2008, pp. 440–444 (2008)
14. Centers for Disease Control and Prevention (CDC): Link Plus (2008). http://www.cdc.gov/cancer/npcr/tools/egistryplus/lp.htm
15. Sariyar, M., Borg, A.: The RecordLinkage package: Detecting errors in data. R J. **2**, 61–67 (2010)
16. MatchWare Technologies Inc: AutoMatch: Generalized record linkage system user's manual (1996)
17. DataLadder: DataMatch. http://dataladder.com/data-matching-software/
18. Campbell, K.M.: The Link King: Record linkage and consolidation software. http://www.the-link-king.com/index.html
19. Campbell, K.M., Deck, D., Krupski, A.: Record linkage software in the public domain: a comparison of Link Plus, the Link King, and a "basic" deterministic algorithm. Health Inform. J. **14**, 5–15 (2008)
20. Gregg, F., Deng, C., Batchkarov, M., Cochrane, J.: usaddress (2014). https://github.com/datamade/usaddress

21. Google Inc.: The Google Maps Geocoding API. https://developers.google.com/maps/documentation/geocoding/intro
22. geocoder.us. http://206.220.230.164
23. Warden, P.: The Data Science Toolkit. http://www.datasciencetoolkit.org
24. Valassis: Residential & Business Database. http://www.valassis.com/direct-mail/mailing-lists/residential-and-business-lists.aspx
25. Winkler, W.E.: The state of record linkage and current research problems. In: Statistical Research Division, US Census Bureau. Citeseer (1999)
26. Jurczyk, P.: FRIL: Fine-grained record integration and linkage tool tutorial, version 3.2 (2009)
27. Cohen, W., Ravikumar, P., Fienberg, S.: SecondString: An open source Java toolkit of approximate string-matching techniques (2003). http://secondstring.sourceforge.net
28. Gisgraphy: Address parser. http://www.gisgraphy.com/
29. Yahoo: Placefinder. https://developer.yahoo.com/boss/geo/
30. Gisgraphy: Gisgraphy results comparison. http://www.gisgraphy.com/compare/
31. Deng, C., Ernsthausen, J.: Parsing addresses with usaddress (2014) Blog article. https://datamade.us/blog/parsing-addresses-with-usaddress
32. Koller, D., Friedman, N.: Probabilistic Graphical Models: Principles and Techniques. MIT Press, Cambridge (2009)
33. Google Inc.: googlemaps 2.2 Python Client for Google Maps Services [21] (2015). https://pypi.python.org/pypi/googlemaps/
34. Yu, X.: pygeocoder 1.2.5. Python interface for Google Geocoding API [21] (2014). https://pypi.python.org/pypi/pygeocoder
35. Choi, S.C.T., Lin, Y.H.: Comparison of Public-Domain Software and Services for Probabilistic Record Linkage and Address Standardization. GitHub repository, https://github.com/schoi32/prl-splncs
36. Szegedy, C., Vanhoucke, V., Ioffe, S., Shlens, J., Wojna, Z.: Rethinking the inception architecture for computer vision. In: Proceedings of the IEEE Conference on Computer Vision and Pattern Recognition, pp. 2818–2826 (2016)
37. Deng, J., Dong, W., Socher, R., Li, L.J., Li, K., Li, F.F.: Imagenet: A large-scale hierarchical image database. In: 2009 IEEE Conference on Computer Vision and Pattern Recognition, CVPR 2009, pp. 248–255. IEEE (2009)
38. Abadi, M., Agarwal, A., Barham, P., Brevdo, E., Chen, Z., Citro, C., Corrado, G.S., Davis, A., Dean, J., Devin, M., et al.: Tensorflow: Large-scale machine learning on heterogeneous distributed systems. arXiv preprint (2016). arXiv:1603.04467
39. Parkhi, O.M., Vedaldi, A., Zisserman, A., et al.: Deep face recognition. In: BMVC, vol. 1, pp. 41.1–41.12. BMVA Press (2015)
40. Vedaldi, A., Lenc, K.: MatConvNet: Convolutional neural networks for Matlab. In: Proceedings of the 23rd ACM international conference on Multimedia, pp. 689–692. ACM (2015)
41. MapQuest: Geocoding API. https://developer.mapquest.com/products/geocoding
42. Microsoft Corporation: Bing Maps REST services. https://msdn.microsoft.com/en-us/library/ff701713.aspx
43. Philips, L.: The double metaphone search algorithm. C/C++ Users J. **18**, 38–43 (2000)
44. NORC at the University of Chicago: Task 4, further PBS research report. Technical report Not published (2012)
45. DataMade: probablepeople. Python library (2014). https://github.com/datamade/probablepeople

Better Interpretable Models for Proteomics Data Analysis Using Rule-Based Mining

Fahrnaz Jayrannejad[1(✉)] and Tim O.F. Conrad[1,2]

[1] Zuse Institute Berlin, Takustr. 7, 14195 Berlin, Germany
jayrannejad@zib.de
[2] Department of Mathematics, Freie Universität Berlin,
Arnimallee 6, Berlin, Germany
conrad@math.fu-berlin.de

Abstract. Recent advances in -omics technology has yielded in large data-sets in many areas of biology, such as mass spectrometry based proteomics. However, analyzing this data is still a challenging task mainly due to the very high dimensionality and high noise content of the data. One of the main objectives of the analysis is the identification of relevant patterns (or features) which can be used for classification of new samples to healthy or diseased. So, a method is required to find easily interpretable models from this data.

To gain the above mentioned goal, we have adapted the disjunctive association rule mining algorithm, TitanicOR, to identify emerging patterns from our mass spectrometry proteomics data-sets. Comparison to five state-of-the-art methods shows that our method is advantageous them in terms of identifying the inter-dependency between the features and the TP-rate and precision of the features selected. We further demonstrate the applicability of our algorithm to one previously published clinical data-set.

Keywords: Bioinformatics · Machine learning · Feature selection · Classification · Association rule mining · Jumping emerging pattern · Proteomics · Mass spectrometry · Clinical data · Biomarker

1 Introduction and Motivation

One of the main objectives when analysing large bio-medical data-sets such as proteomics data (or other large omics data-sets, such as genomics) is to derive classification models that allow distinguishing between phenotypes, e.g. healthy and diseased samples. Finding a good classifier is a well studied problem in the area of supervised learning. However, most machine learning approaches solving this problem often result in very complex (often non-linear) models that offer very high classification accuracy but are very hard to interpret. A better approach would be a model that offers high accuracy and - at the same time - allows derivation of simple to understand rules explaining *why* the two classes are different. These rules can then be used to gain a deeper understanding of the

A. Holzinger et al. (Eds.): Integrative Machine Learning, LNAI 10344, pp. 67–88, 2017.
https://doi.org/10.1007/978-3-319-69775-8_4

Table 1. Example 1

	Sex	Age	Mutation	History	Smoking	Risk
Case 1	Male	39	Yes	Yes	Yes	**High**
Case 2	Male	72	Yes	No	No	**High**
⋮	⋮	⋮	⋮	⋮	⋮	⋮
Case n	Male	45	Yes	Yes	No	**High**
Control 1	Female	23	No	No	Yes	**Low**
Control 2	Male	22	No	No	No	**Low**
⋮	⋮	⋮	⋮	⋮	⋮	⋮
Control m	Female	22	No	No	No	**Low**

Table 2. Results for Example 1: State of the art methods

Method	Model	Features
NMF	–	Sex, mutation, smoking
Lasso	$-1.0 - 0.8*$Smoking $+ 1.0*$history $+ 1.2*$Mutation $+ 0.003*$Age $+ 0.2*$Sex	Sex, mutation, smoking, history, age
SVM	$-0.9 - 0.3*$Smoking $+ 0.6*$history $+ 1.0*$Mutation $+ 0.004*$Age $+ 0.3*$Sex	Sex, mutation, smoking, history, age
Rpart	if (Age>=34) then **High Risk** if (Age<=34) then **Low Risk**	Age
JRip	if (Age>=39) then **High Risk** if (Age<=39) then **Low Risk**	Age

Titanic	if ((Age>34) or (Mutation=Yes)) then **High Risk**
	 if ((Age>34) or (History=Yes)) then **High Risk**
	 if ((History=Yes) or (Mutation=Yes)) then **High Risk**
	 (if (Sex=Female) or (Age<34)) then **Low Risk**
		Sex, Mutation, History, Age

problem and the data at hand. Let us consider the following example: Table 1 shows a hypothetical data-set that could come from some cancer study. The goal is to find a model (classifier) that can be used to predict the risk of an unknown sample.

Now let us take a look at the results of some widely used methods (Table 2), such as SVM [1–3] or Lasso [4,5]. These methods usually result in a linear combination of the most important features. However, these models are not easily interpretable and they do not give any information regarding the inter-dependency between the features. On the other hand, rule-based methods such as Rpart [6], JRip [7] or TitanicOR [8] result in a set of easily interpretable rules, which can even show inter-dependencies between features.

Decision trees also provide readable association rules as the classification model. But, they are different from ours. Some of the major differences are as follows:

Decision trees have a hierarchical model, whilst ours provides general rules. Decision trees select only features which their values in each class obey a specific distribution. But, ours even selects the features with specific distribution in only one class. Decision trees select either a group of features which have association rule together, or only one single feature which could distinguish the two classes. Ours selects all the association rules and single distinguishing features. Due to the heuristic manner of decision trees, most of the association rules and single distinguishing features are neglected. Our method selects all of the distinguishing association rules and single features.

In recent years, many methods have been proposed for solving the feature selection problem, e.g. SVM (L1 regularised), Lasso and Binda [9]. Table 3 provides a brief overview of the current ML methods applicable to this problem.

One of the feature selection methods which performed particularly well is Binda. Binda assigns a score to each feature and ranks them based on their scores. Later, the classifier tries to find the least number of features need to get the accurate classification.

It should be noted that we focus in this paper only on methods from the area of *automatic* machine learning where the task (e.g. feature selection) is solved only by an algorithm without human interaction (also called "human-out-of-the-loop"). However, recent advances in the area of *interactive* machine learning (also called "human-in-a-loop") [10,11] show exciting results in analysing datasets where usually neither human nor computer algorithms could solve on their own.

Table 3. A brief comparison of the current state-of-the-art ML algorithms

Algorithm	Complexity of the Model	Features	Interdependency of the features
SVM	High	Single Features List	No
Lasso	High	Single Features List	No
Binda	High	Features Ranking	No
Decision Trees	Low	Conjunctive Rules	Yes
Group Lasso [12]	High	Features	Yes
NMF [13]	High	Features	Yes
ZBDD [14]	Average	CNF Rules	Yes
Disclosed [15,16]	Average	CNF Rules	Yes
BLOSOM [17]	Low	Disjunctive minimal generators	Yes
QCEP [15]	Average	CNF Rules	Yes
TitanicOR [8]	Low	Disjunctive minimal generators	Yes

Many studies have been done to evaluate the performance of different Feature selection and classification algorithms on MALDI-MS data. Here, we only discuss the results of two surveys which have tested most of the state-of-the-art algorithms on MALDI-MS data.

In 2009, Liu et al. [18], tested a group of feature selection and classification algorithms on MALDI-MS data. In this study, SVM outperformed most of the other methods.

In 2013, Swan et al. [19], did a survey on the application of different machine learning methods in MS-Proteomics data analysis, which showed that Naive Bayes was the fastest algorithm and SVM and ANN were the slowest ones, whilst Decision tree and Rule-based classifiers developed the most easily interpretable models and SVM and ANN (Artificial Neural Networks) lead in the most complex models.

The speed of the classification is not the main concern in biological data analysis. Here, the main concern is to find the most relevant features, build a classifier with the highest possible accuracy and provide a model easily understandable to the end users, e.g. biologists. Therefore, we are interested in applying *association rule mining* algorithms to get such classification models. Now lets get a quick look at the literature of *association rule mining*.

Association Rule (AR) Mining

Association rule mining algorithms have been designed to identify correlation relationships between variables in a given data-set. They became popular due to the 1993 article of Agrawal et al. [20]. The method suggested here is to use association rule mining algorithms to identify discriminating features from mass spectrometry (MS) data. Many other techniques, including machine learning methods, have been proposed to solve this problem. One of the purposes of proteomics studies is to find features which can be used in the diagnosis of diseases and provide an easily interpretable model. On the other hand, most diseases are not considered to be connected to a single cause (or feature), rather they are connected to a group of features together [16,21,22]. Association rules can show this inter-dependency in an easily understandable and interpretable way.

Our main concern in current study is to provide an easily readable and understandable classifier model to the users and identify the interdependency between the features. Decision trees, BLOSOM and TitanicOR are the only methods in Table 3, which satisfy both of these criteria at the same time. As discussed earlier, decision trees due to their heuristic manner, which tends to select the minimum number of the features, neglect a big part of the information which might be interesting to the doctors. Studies [23] have proven that minimal generators outperform closeditem-sets (both discussed later in Sect. 3.2) in classification problems. BLOSOM is a state-of-the-art algorithm in mining disjunctive minimal generators. Later, Vimieiro et al., introduced TitanicOR which despite the depth-first search manner of BLOSOM does a breadth-first search and has outperformed BLOSOM in application to many data-sets [16].

This is a preliminary study to show the applicability of this kind of method and this will be further investigated with more data.

The rest of this paper is organised as follows:

In Sect. 1.1., there is the problem definition, followed by data description in Sect. 4. Our methodology and preliminary concepts are explained using the example from Sect. 1.1 in Sect. 3.

The experimental results are discussed in Sect. 5 and finally, we conclude our paper in Sect. 6.

1.1 Problem Definition

As discussed, one of the main purposes of mass spectrometry studies is to find biomarkers from the MS data which are associated with the case of study, e.g. a particular disease.

Considering the examples in Fig. 1 as a sample spectrum, the m/z values having high intensity could be potential biomarkers.

One of the challenges here is to distinguish the peaks which have different intensities in different classes. Like the ones specified in Fig. 2. In these two examples, the difference of the intensities is easily distinguishable by eye. In most of the cases, this is not possible and they don't differ as much as the samples in Fig. 2. On the other hand, in real studies, we are dealing with a set of hundreds of samples, each having hundreds of peaks. So, we need a method which can identify the m/z values in which the intensities in different classes are different and find out how are they different. As an example, in Fig. 2, the intensity at 1866.18 is 0.0004 in one sample and 0.0001 in the other one. Thus, it is probable that there is a threshold for the intensity at 1866.18 which could classify the samples into two classes.

Our main concern here is, to identify the distinguishing m/z values and the threshold associated to them which can classify the samples. As discussed in Sect. 1, this m/z values should be given in an easily interpretable way for the end users. The interdependency between the m/z values is another interesting information which we are looking for.

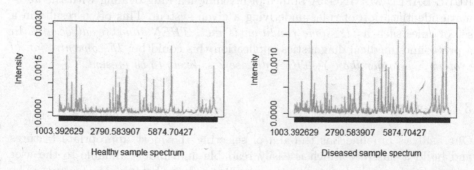

Fig. 1. Two sample spectra (It is difficult to identify the peaks from the raw data by eye, so here we use the intensities from the pre-processing discussed in Sect. 3).

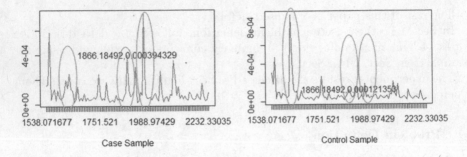

Fig. 2. A part of the samples spectra in Fig. 1 and the potential biomarkers (It is difficult to identify the peaks from the raw data by eye, so here we use the intensities from the pre-processing discussed in Sect. 3).

2   Glossary

PROTEOMICS: The goal of proteomics is to obtain a global and integrated understanding of biology by studying all proteins (molecules) in an organism rather than individual proteins. Proteins are the *working horses* in most living organisms. One example application of proteomics is the comparison of biological samples of two different conditions, such as healthy individuals and patients having a particular disease. The main goal is then to identify differences on a proteomic level which can serve as biomarkers in new types of disease diagnostics.

MASS SPECTROMETRY: Mass Spectrometry (MS) is an technology for the analysis of (biological) samples. It works by first ionizing the molecules (e.g. proteins) in a given sample and then measures their masses (or better: their mass-to-charge ratio). The result of this process is often a mass spectrum plot showing how many molecules (ions) of a particular mass have been contained in the given sample.

RULE BASED MINING: A branch of machine learning dealing with the automatic identification of rules underlying a given system. This often results in a set of rules such as *IF some condition is met, THEN some result occurs..* In a proteomics medical diagnostics application this could be: *IF concentration of protein X is larger than Y, THEN disease Z is likely to be present.*

3   Methods

Our analysis pipeline has the aim of selecting the most appropriate features and build a classifier which is easily readable and understandable to the user and gives information regarding the inter-dependency between the features. The main steps are shown in Fig. 3. We will now explain the individual steps.

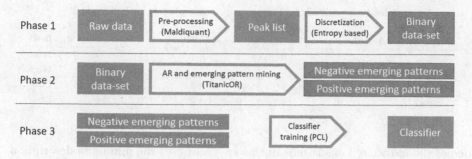

Fig. 3. The proposed pipeline. The raw data is fed to Maldiquant to do the pre-processing and get the peak list from the data-set. Later, the intensity values from the pre-processing will be discretized using an entropy-based algorithm. TitanicOR will mine the association rules and emerging patterns from the discretized data. PCL will classify new unlabeled samples using the emerging patterns from the previous step.

3.1 Preprocessing

Peak Detection. The original raw MS data has more than thousands of mass values. But only the peaks have some information that can help us to classify the samples. Therefore, a preprocessing should be applied to the raw data. Here, we use Maldiquant [24] to extract the peaks. Peak extraction usually is done after some preprocessing steps to standardise the data. We used the following preprocessing steps:

(1) Variance Stabilisation using the square root transformation to simplify graphical visualisation and to overcome the potential dependency of the variance from the mean. (2) Smoothing using a 21 point Savitzky-Golay-Filter [25] to smooth the spectra. (3) Intensity Calibration/Normalisation using the Total-Ion-Current-Calibration for better comparison and to overcome (very) small batch effects we equalise the intensity values. (4) Warping/Alignment using a peak based warping algorithm for (re)calibrating the mass values [26]. (5) Peak detection using windowing.

Discretization. As discussed in Sect. 1, the main core of our suggested method is the association rule mining algorithm. Association rule mining algorithms work only with binary-valued features data-sets (transactional data-sets). In Table 1, sex is a categorical feature and age is integer valued, whilst the rest are binary-valued (the value "Yes" is replaced by "1" and the value "No" is replaced by "0"). Thus, the first two features need to be mapped to binary-valued features. The feature "Sex" can be mapped into three other binary-valued features "Sex=Male", "Sex=Female" and "Sex=Other". The feature "Age" is integer valued, therefore some preprocessing should be applied to it to get binary-valued features from it, it is the same about the features in MALDI-MS based data-sets and the same preprocessing is applied to them. Here, it will be converted to a categorical feature using the method suggested by Irani et al. [27,28], which is

Table 4. A part of the binary-valued data-set from Example 1

	Sex=Female	Sex=Male	Age<34	Age>34	Mutation	History	Smoking	Risk
Case1	0	1	0	1	1	1	1	**High**
Case2	0	1	0	1	1	0	0	**High**
Control1	1	0	1	0	0	0	1	**Low**
Control2	0	1	1	0	0	0	0	**Low**

one of the most cited algorithms in this area and uses the minimum description length principle [29].

Later the categorical values can easily be converted to binary-valued features. Here, based on the minimum description length the threshold 34 is selected to convert the feature *"Age"* into a categorical feature with the minimum error. Thus, it is converted to a categorical feature with two possible values "Age < 34" and "Age > 34", so the feature age will be replaced with two binary-valued features "Age < 34" and "Age > 34".

So the data-set will be converted to a data-set with binary-valued features, like the sample data-set in Table 4.

3.2   Feature Selection

The association rule mining algorithm will be applied to binary-valued data-sets of high risk and low risk separately. The output is a list of features and association rules for each class, which their frequency in one class is higher than a minimum threshold (α) and less than a maximum threshold in the opposite class (β). The thresholds (α) and (β) are set based on the data-set and it's density.

The proposed method is to use association rule described in Sect. 3.2 and jumping emerging pattern, Definition 5, mining algorithms to find appropriate classifying features from the peak list retrieved from the input data-sets [24].

TitanicOR is the association rule mining algorithm selected here to mine the association rules and emerging patterns from them. Later, PCL (Prediction by Collective Likelihood), will be used to classify the new samples into healthy and diseased.

In order to understand TitanicOR, we need to go through some concepts.

Definition 1 (Data-Set). *Let S be a set of sample IDs, I a set of items and $R \subseteq S \times I$ be a binary (incidence) relation, where $(s,i) \in R$ shall be read as 'the sample s has the item i'. A data-set D is the triplet (S, I, R)* [8].

Definition 2 (Frequentitem-set). *A subset Y of I is called anitem-set. Let $f : P(I) \rightarrow P(S), f(Y) = \{s \in S | \exists i \in Y[(s,i) \in R]\}$ denote the set of samples associated with the items in Y* [16] *- we use $P(X)$ to denote $W \subseteq X$. We call $|f(Y)|$ the (disjunctive) support of anitem-set Y. We say that anitem-set is frequent if its support is at least as great as a user specified minimum support threshold α* [8].

Definition 3 (Frequent Closeditem-set). *An item-set X is closed if none of its immediate supersets has exactly the same support count as X.*

Definition 4 (Minimal Generator). *A set $X \subseteq M$ is a key set (or minimal generator) if X is minimal (with respect to set inclusion) in $[X]$ [30].*

3.3 TitanicOR

The TitanicOR [8] algorithm is an adaptation of the original conjunctive version of Titanic proposed by Stumme et al. [30] that computes disjunctive association rules from binary data-sets (see Definition 1) with frequency higher than a pre-specified threshold (α) and less than a threshold (γ) which will be set based on the data-set and it's density.

TitanicOR operates in a "level-wise" manner like Apriori [31]; it first finds all the minimal generators (Definition 4) of size k before it starts computing item-sets of size k + 1.

3.4 Prediction by Collective Likelihood (PCL)

PCL [32] is based on the concept of jumping emerging patterns [33].

Definition 5 (Emerging Pattern). *Given a positive data-set D_p, a negative data-set D_n, and support thresholds α and β. An Emerging Pattern (EP) [14] is an item-set (see Definition 2) p satisfying two support constraints, (i) $support(p, D_n) \leq \beta$ and (ii) $support(p, D_p) \geq \alpha$. Furthermore, p is a minimal EP if p does not contain any other item-set that satisfies constraints i-ii.*

Definition 6 (Jumping Emerging Pattern). *A jumping emerging pattern is an emerging pattern which the threshold (β) is set to zero.*

In other words, an emerging pattern is a set of conditions often including several features, with which most of a class of samples' expression satisfy, but none of other class's samples satisfies. So, an emerging pattern can be considered as a multi-feature discriminator. The central spirit of PCL is to use top-ranked multi-feature discriminators to make a collective prediction. PCL uses feature groups, does not assume that features are independent; PCL can provide more than a mere prediction or a distance, but many interesting rules.

Given two training data-sets D_p and D_N and a testing sample T, the first phase of the PCL classifier is to discover EPs from D_p and D_N- Denote the ranked EPs of D_p as,

$$EP_1^{(P)}, EP_2^{(P)}, EP_3^{(P)}, ..., EP_i^{(P)}$$

in descending order of their frequency. Similarly, denote the ranked EPs of D_N as

$$EP_1^{(N)}, EP_2^{(N)}, EP_3^{(N)}, ..., EP_j^{(N)}$$

also in descending order of their frequency.

Suppose $\acute{T}$ contains the following EPs of D_p:

$EP_{i1}^{(P)}, EP_{i2}^{(P)}, EP_{i3}^{(P)}, ..., EP_{ix}^{(P)}$, where $i1 < i2 < \cdots < ix \leq i$, and the following EPs of D_N:

$EP_{j1}^{(N)}, EP_{j2}^{(N)}, EP_{j3}^{(N)}, ..., EP_{jy}^{(N)}$, where $j1 < j2 < \cdots < jy \leq j$.

The next step is to calculate two scores for predicting the class label of T. Suppose we use k ($k \leq i$ and $k \leq j$) top-ranked EPs of D_p and D_N. Then we define the score of T in the D_p class as

$$Score(T)\_D_P = sum_{K=1}^{m} \frac{frequency(EP_i m^{(P)})}{frequency(EP_m^{(P)})} \tag{1}$$

and similarly the score in the D_N class as

$$Score(T)\_D_N = \sum_{K=1}^{m} \frac{frequency(EP_i m^{(N)})}{frequency(EP_m^{(N)})} \tag{2}$$

If $Score(T)\_D_P < Score(T)\_D_N$, then T is predicted as the class of D_N. Otherwise it is predicted as the class of D_P.

As discussed, the algorithm TitanicOR extracts the frequent disjunctive itemsets from the data of each class. Here, after applying steps mentioned in Sect. 3, we apply TitanicOR on each class of data separately and extract the emerging patterns from the association rules generated by TitanicOR.

Later Top k emerging patterns from each class were selected to do the classification based on them. The number k depends on the data-set and should be selected based on the problem.

4   Data

In this study, we conducted our experiments on two sets of data. A simulated data-set to show analyse the behaviour of the tested algorithms with a known ground-truth. To also evaluate the methods on real data we perform the analysis on previously published, publicly available data-sets.

4.1   Simulated Data

A proper method should be able to identify peaks with different intensities, so five groups of data-sets with different intensities were simulated [24], which are divided into two groups (same as the real life problems which are divided into two classes as diseased and healthy, here called as case and control) with the same number of samples. The real ms samples normally measure the intensity of tens thousands mass values and after doing pre-processing which will be discussed later in Sect. 3.1 there will be a list of hundreds of peaks extracted from them. Therefore, we simulated spectra with more than 40 thousands of mass values and 416 peaks in each group which are selected as follows. There are 400 peaks

Table 5. The distribution of the peaks in their corresponding groups, in the simulated data-sets. The peaks from class case are specified in red and the ones from control are green. We are interested in evaluating the performance of our method in identifying the features and association rules between the features. Therefore, we divided the samples from case to three groups A, B and C. Sixty percent of the samples are in group A, twenty percent are in group B and twenty percent in group C. Each group contains 12 peaks out of the 16 peaks from case. The + signs in the table indicate each peak is present in which group. The column frequency shows the frequency of each peak in the samples from case. We repeated the same procedure with the same settings for the samples from control class.

Peaks	Case (A, 60%)	Case (B, 20%)	Case (C, 20%)	Control (A, 60%)	Control (B, 20%)	Control (C, 20%)	Frequency
1501	+	+					80
1647				+	+		80
1800						+	20
1960			+				20
2127				+		+	80
2301	+		+				80
2481					+		20
2668		+					20
2862		+					20
3063				+	+		80
3271			+				20
3485	+		+				80
3706			+				20
3934					+		20
4169	+	+					80
4411				+		+	80
4660				+	+	+	100
4915	+	+	+				100
5177	+	+	+				100
5446				+	+	+	100
5722	+	+	+				100
6005				+	+	+	100
6294				+	+	+	100
6590	+	+	+				100
6893				+	+	+	100
7203	+	+	+				100
7520				+	+	+	100
7843	+	+	+				100
8174	+	+	+				100
8511				+	+	+	100
8855	+	+	+				100
9206				+	+	+	100

in fixed positions in both groups. In addition, there are 16 extra peaks in each group. Table 5 shows how the peaks were distributed in the samples of each class.

In order to simulate the spectra, for each spectrum, we first generated a list of random intensities with the distribution $N(500, 1)$. Then for each selected peak we added a random intensity from the distribution specified below each table (e.g. $N(15000, 1500)$) to the selected samples from it's corresponding class and an intensity from normal distribution $N(100, 1)$ to the rest of the samples from it's corresponding class and all the samples from the opposite class.

4.2 Real Data

This data is from a study conducted by Fiedler et al. [34] in 2009. For the discovery study, two sets of sera from patients with pancreatic cancer ($n = 40$) and healthy controls ($n = 40$) were obtained from two different clinical centres. For external data validation, we collected an independent set of samples from patients ($n = 20$) and healthy controls ($n = 20$) [35].

5 Experimental Results

The pipeline discussed in Sect. 3 was applied on two different kinds of data-sets (simulated and real data-set) to measure the accuracy of the classifier and the relevance of the features selected.

The raw data is fed to the Maldiquant to do the pre-processing and get the peak list from the data-set. Later, the intensity values from the pre-processing will be discretized using an entropy-based algorithm. TitanicOR will mine the association rules and emerging patterns from the discretized data. PCL will classify new unlabeled samples using the emerging patterns from the previous step.

Six other algorithms were selected to be compared with the proposed pipeline. SVM is the state-of-the-art algorithm, successful in most classifications. Binda is a package specially implemented for binary classification which has performed well on MS proteomics data-sets. Lasso is one of the powerful feature selection and regression (classification) algorithms. Rpart and Jrip are two rule-based classifiers which select the least number of the powerful features need for classification. Rpart reports a number of top features, here 6, and builds the classification rules based on them. Here, classification is possible using only one rules made of one feature. So, all of the association rules between the features and most of the single features are neglected.

NMF is a sparse feature selection and classification algorithm.

We tried different thresholds to set the experiment. The best results were obtained using the following thresholds:

$\alpha = 80\%$, $\beta = 0\%$, $k = 5$, $\gamma = 100\%$ (in MALDI-MS data most of the important features appear in all of the samples from their corresponding class).

5.1   Results on the Simulated Data-Set

Considering the simulation settings given in Table 5, the following single features and association rules should be selected.

Table 6. The features and association rules we expect to be selected based on the simulation settings. The features and association rules from different groups are shown by different colours. The upper half of the table shows the inter-dependency between the features which has been discussed in Sect. 1.1. The lower part shows the single features which have occurred in all of the samples from their corresponding class. T hese features do not have any association with any other feature.

1501 or 2301	2301 or 4169	1501 or 3458	3485 or 4169
1647 or 2127	2127 or 3063	3063 or 4411	1647 or 4411

4915	5177	5722	6590	7203	7843	8174	8855
4660	5446	6005	6294	6893	7520	8511	9206

The single features listed in Table 6 (lower part) are the single features which have occurred in all of the samples from their corresponding class, these features do not have any association with any other feature. They are only associated with the class label. The other ones (upper part) are the association rules between the features with frequency 80% in their corresponding class. The features with frequency 20%, are representing the features which cause other phenotypes to occur and are not related to our classification problem, e.g. diabetes might cause obesity, so some of the cases diagnosed with diabetes might have the peptide responsible for obesity as well, but those peptides are not important in classifying samples diagnosed with diabetes. Therefore, we are expecting 24 features to be selected by each method, either in association with other feature or as single features.

The results of the simulated data-sets are summarised in Tables 7, 8, 9, 10, 11, 12 and 13.

Here, as a metric for comparing different methods, we compare the TP-rate and precision of the features selected.

Binda assigns a score to each feature and ranks them based on their scores. Therefore, here we just get the first 24 features based on the binda ranking to

Table 7. Results from Simulation N(15000,1500)

	SVM	Binda	Lasso	Rpart	Jrip	NMF+ SVM	TitanicOR
*Accuracy**	1	1	1	1	1	1	1
#ofFeatures	29	24	16	6	1	24	24
#ofTPFeatures	13	24	16	6	1	24	24
#ofAssociationRules	–	–	–	–	·–	–	8
TP − rate	0.54	1	0.67	0.25	0.04	1	1
Precision	0.44	1	1	1	1	1	1

Table 8. Results from Simulation N(10000,1000)

	SVM	vBinda	Lasso	Rpart	Jrip	NMF + SVM	TitanicOR
Accuracy*	1	1	1	1	1	1	1
#ofFeatures	18	24	16	6	1	24	24
#ofTPFeatures	14	24	16	6	1	24	24
#ofAssociationRules	–	–	–	–	–	–	8
TP − rate	0.75	1	0.67	0.25	0.04	1	1
Precision	0.77	1	1	1	1	1	1

Table 9. Results from Simulation N(5000,500)

	SVM	Binda	Lasso	Rpart	Jrip	NMF + SVM	TitanicOR
Accuracy*	1	1	1	1	1	1	1
#ofFeatures	12	24	16	6	1	24	24
#ofTPFeatures	9	24	16	6	1	23	24
#ofAssociationRules	–	–	–	–	−2	–	8
TP − rate	0.37	1	0.67	0.25	0.04	0.96	1
Precision	0.75	1	1	1	1	0.96	1

Table 10. Results from Simulation N(2000,200)

	SVM	Binda	Lasso	Rpart	Jrip	NMF + SVM	TitanicOR
Accuracy*	1	1	1	1	1	1	1
#ofFeatures	20	24	16	6	1	23	24
#ofTPFeatures	19	24	16	6	1	23	24
#ofAssociationRules	–	–	–	–	–	–	8
TP − rate	0.79	1	0.67	0.25	0.04	0.96	1
Precision	0.95	1	1	1	1	1	1

Table 11. Results from Simulation N(1000,100)

	SVM	Binda	Lasso	Rpart	Jrip	NMF + SVM	TitanicOR
Accuracy*	1	1	1	1	0.99	1	1
#ofFeatures	19	24	17	6	1	24	24
#ofTPFeatures	17	24	17	6	1	22	24
#ofAssociationRules	–	–	–	–	–	–	8
TP − rate	0.7	1	0.7	0.25	0.04	0.91	1
Precision	0.89	1	1	1	1	0.91	1

* Here Accuracy means how accurate the algorithms could classify the samples. The row number of Association Rules shows the number of the association rules reported between the features.

evaluate the performance of Binda, since there are 24 discriminating features in each data-set.

NMF has a random initialization and doesn't give stable results. Thus, we ran it 100 times and later selected top 24 features (two sets of 12 features) based on the average score of each run.

This data is linear classifiable and one of the main challenges here is information retrieval. Thus, here we concentrate more on the features selected and how much information we can get from them.

As discussed earlier in Sect. 3.4 and this Sect., there are 24 features which are present only in one class with a high frequency, so we consider them as distinguishing features and expect them to be among the selected features.

Tables 7, 8, 9, 10, and 11 show that among all seven ML methods applied on the data-sets, our method and Binda are the most powerful ones. Our method selected only the 24 discriminating features. There are 24 features from the expected features among top 24 ranked by Binda.

An other interesting information which we look for is the inter-dependency between the features (not between single features and the class labels). Rpart and Jrip are two AR based algorithms which we have applied to the data-sets. As discussed, this data is linear classifiable and can be classified even with one or two features. So, even Rpart and Jrip failed to distinguish the inter-dependency between the features. They only reported ARs between single features and class labels (e.g. if $intensity_{4660}$ is greater than a threshold then Class = 'Control').

Our method succeeded in finding the inter-dependency between the features and the class labels as well, which has interesting information for the users e.g. medical doctors about how proteins and peptides interact with each other.

The rest of the algorithms didn't find any AR.

The ARs from Table 7 is given in Table 12.

Table 12. Selected association rules from Table 7. The red peaks are the peaks we added in the case samples group and the blue ones are the peaks added in the control samples group.

Item-sets from class "Control"	Item-sets from class "Case"
$intensity_{1647} > 12059.04$ or $intensity_{2127} > 603.38$	$intensity_{1501} > 12269.76$ or $intensity_{2301} > 602.57$
$intensity_{1647} > 12059.04$ or $intensity_{4411} > 12846.62$	$intensity_{2301} > 602.57$ or $intensity_{4169} > 9309.87$
$intensity_{3063} > 12270.38$ or $intensity_{4411} > 12846.62$	$intensity_{3485} > 11133.89$ or $intensity_{4169} > 9309.87$
$intensity_{4460} > 11735.97$	$intensity_{4460} < 11735.97$
$intensity_{4915} < 11770.84$	$intensity_{4915} > 11770.84$
$intensity_{5177} < 12044.52$	$intensity_{5177} > 12044.52$
$intensity_{5446} > 10555.56$	$intensity_{5446} < 10555.56$
$intensity_{5722} < 11603.59$	$intensity_{5722} > 11603.59$
$intensity_{6005} > 12699.46$	$intensity_{6005} < 12699.46$
$intensity_{6294} > 12122.54$	$intensity_{6294} < 12122.54$

Table 13. Rules from Table 12. The ¬ (not) sign indicates a negative association rule. The blue colour indicates peaks appearing in the control class and red colour indicates the peaks from other (case) class. The top 4 rows from the features selected for each class show the association (inter-dependency) between the features. As shown in Table 6, we expect from each group 4 association rules (inter-dependencies between the features) which are the same as the association rules listed here.

Rules from class Control	Support(%)	Rules from class Case	Support(%)
$Peak_{1647}$ or $Peak_{2127}$	100	$Peak_{1501}$ or $Peak_{2301}$	100
$Peak_{2127}$ or $Peak_{3063}$	100	$Peak_{1501}$ or $Peak_{3271}$	100
$Peak_{1647}$ or $Peak_{4411}$	100	$Peak_{2301}$ or $Peak_{4169}$	100
$Peak_{3063}$ or $Peak_{4411}$	100	$Peak_{3271}$ or $Peak_{4169}$	100
$Peak_{4460}$	100	$\neg Peak_{4460}$	100
$\neg Peak_{4915}$	100	$Peak_{4915}$	100
$\neg Peak_{5177}$	100	$Peak_{5177}$	100
$Peak_{5446}$	100	$\neg Peak_{5446}$	100
$\neg Peak_{5722}$	100	$Peak_{5722}$	100
$Peak_{6005}$	100	$\neg Peak_{6005}$	100
$Peak_{6294}$	100	$\neg Peak_{6294}$	100
$\neg Peak_{6590}$	100	$Peak_{6590}$	100
$Peak_{6893}$	100	$\neg Peak_{6893}$	100
$\neg Peak_{7203}$	100	$Peak_{7203}$	100
$Peak_{7520}$	100	$\neg Peak_{7520}$	100
$\neg Peak_{7843}$	100	$Peak_{7843}$	100
$\neg Peak_{8174}$	100	$Peak_{8174}$	100
$Peak_{8511}$	100	$\neg Peak_{8511}$	100
$\neg Peak_{8855}$	100	$Peak_{8855}$	100
$Peak_{9206}$	100	$\neg Peak_{9206}$	100

The threshold selected in the discretization phase described in Sect. 3, can also be considered as a threshold to indicate if the peak appears in the sample or not. Therefore, Table 12 can be represented as in Table 13.

The features and association rules show in Table 13 will be used as input for the PCL algorithm (see Definition 6), to do the classification. As shown in Table 13, negative association rules are reported for the features which cover all the samples from their corresponding class and none from the opposite class. However, there is no negative association rule for other features. Usually, there are peptides (and thus peaks) in mass spectrometry data for which their intensities in opposite classes follow different distributions. Therefore, those features will be selected as discriminating features which occur in all of the samples in one class and none of the samples in the opposite class. Thus, the support reported for them will be (100%) in their corresponding class and all of them have the same importance in the classification. When any new sample is checked to be classified using the associations in Table 13 and PCL, most probably it will have the features only one class. Any new sample could be classified using the ARs from Table 13 with high accuracy, e.g. $Peak_{1647}$ *or* $Peak_{4411}$ *could be interpreted as:* "**IF** *there is a peak at either m/z value 1647 or 4411* **THEN** *the sample belongs to class control*". The $\neg Peak_{9206}$ *rule could be interpreted as:* "**IF** *there is no peak at m/z value 9260* **THEN** *the sample belongs to class case*". Of course, the samples can be classified using the single features. The main importance of the association rules from the first four rows of this table is to understand the relation and inter-dependency of the features and peptides with each other better. This could help the biologist to get a more precise idea on the causes of phenotypes and the way they could cure or prevent a phenotype from occurring.

As shown in Tables 12 and 13, the rules from TitanicOR have the following advantages: (1) The rules are easily readable. (2) Each rule has information regarding the intensity at a certain mass value in the corresponding class. (3) Considering the thresholds selected by discretization method described in Sect. 3 as the threshold for the presence or absence of each peak, we can also have negative association rules from the data-sets. As an example, $\neg Peak_{5446}$ can be interpreted as *"There is not a peak at m/z value 5446"*. This kind of interpretation is not possible for models from other methods. In the simulated data-sets, the peaks in Table 5 do not have the same frequency in their corresponding classes. Four out of sixteen peaks of each class occur in 20% of the simulated samples, four occur in 80% of the samples (20% of this 80% overlaps with the 20% samples of the four peaks from the second group) and the rest are present in all of the samples. The first group of the peaks represent the features which might be present in a group of the samples but are not related to the disease under study, e.g. in some cases, one disease might cause another disease to happen, so the peptides for the later disease might occur in some cases of the first one, but they can not help in distinguishing it.

After applying our method on the data-sets, all of the peaks from the third group were selected and the support reported for them was 100%. The other peaks with 80% were selected as well. The peaks from the first group (with frequency 20%) were not selected.

As discussed, there is overlap between the samples containing each group of peaks, therefore it is expected to get some association rules between the features. As reported in the tables, some association rules between the single features and the class labels and item-sets of the single features and the class labels were detected. This item-sets show the inter-dependency (association rule) between the features.

All the features and the item-sets selected had the support expected (respect to the simulation settings).

One concern about the ARs selected by our algorithm is if some irrelevant ARs are selected. Here, there was no irrelevant feature reported. Therefore, all of the ARs selected by our algorithm are between features related to our classification problem. Also, all the association rules selected were the same as the association rules we expected to be selected.

5.2   Results on the Real Data-Set

Finally, the pipeline was applied to a real data-set from a published study [34] (Table 14).

Table 14. Results from real data-set

	SVM	Binda	Lasso	Rpart	Jrip	NMF + SVM	TitanicOR
*Accuracy**	0.93	0.9	0.94	0.87	0.97	0.9	0.9
#ofFeatures	23	–	34	6	1	44	10
#ofAssociationRules	–	–	–	–	–	–	4

The same arguments as in Sect. 5 regarding the classification and information retrieval and the inter-dependency between the features and understandability of the model apply here as well. None of the algorithms reported any inter-dependency, Although Rpart and Jrip are two rule-based algorithms, they were not able to find the inter-dependency between the features and as discussed in the experimental results from the simulated data-sets they only identified two association rules between the value of one feature and the class label.

This inter-dependency could show how the proteins and peptides might affect each other and which peptides are related to each other.

TitanicOR could do the classification using even one of the features reported, but algorithms such as SVM need to use all of the features.

One (any body even without any knowledge of data science) could interpret the rules reported by TitanicOR, but the models from the rest are not easily understandable by users specially the users from medical science who are not familiar with data science.

As discussed earlier in Sect. 1, Binda assigns a score to each feature and ranks the features based on those scores, there is not a list of discriminating features like what we get from the other algorithms. It has also performed very well on MALDI-MS data. As an evaluation of the features selected by TitanicOR and Binda, we compared top 10 features selected by Binda. All of the top 10 ranked features by Binda were also selected by TitanicOR. This overlap suggests the relevance of the features selected by both methods. In order to evaluate the features selected by TitanicOR, we compared them with the features selected by other methods in Table 15.

Table 15. Number of the features that are common between the features selected by TitanicOR and each of the other algorithms

Algorithm	SVM	Binda	Lasso	Jrip	Rpart	NMF
# of features in common with TitanicOR	5	10	3	1	6	6

Jrip and Rpart only reported respectively first and 6 important features from the features. In other words, they are trying to identify the minimal features need to do the classification, which may lead in neglecting many features which might be interesting for the end users, e.g. medical doctors.

Table 16 shows the association rules selected from the real data-set, the peaks in red and blue show the peaks which obey a certain distribution in each class. In other words, we have information regarding their absence or presence in each class. The peaks in black show the peaks which we have information regarding their absence or presence in only one class, e.g. we have $Peak_{4442}$ for class case and $\neg Peak_{4442}$ for class control, which could be interpreted as *"if a sample has a peak at m/z value 4442, then it should be associated with class case, otherwise it is from class control"*. There is also *"$Peak_{5906}$"* for class control, which could be interpreted as *"if there is a peak at m/z vale 5906, then it is healthy, but if there is no peak at this m/z value, we can not conclude anything"*. Considering *"$Peak_{2022}$ or $Peak_{8867}$"*, there are four possibilities, (1) there are peaks at both m/z values 2022 and 8867, (2) there is only peak at m/z value 2022, (3) there is peak only at m/z value 8867,(4) there is no peak at neither m/z value 2022, nor m/z value 8867. If any of the cases (1)–(3) happens then that sample is diseased, otherwise it is healthy. So, if there is peak at either m/z values 2022 or 8867, shows the occurrence of the target phenotype.

Table 16. The association rules from Table 15, the peaks in red and blue show the peaks which obey a certain distribution in each class. In other words, we have information regarding their absence or presence in each class. The peaks in black show the peaks which we have information regarding their absence or presence in only one class, e.g. we have $Peak_{4442}$ for class case and $\neg Peak_{4442}$ for class control, which could be interpreted as *"if a sample has a peak at m/z value 4442, then it should be associated with class case, otherwise it is from class control"*. There is also *"$Peak_{5906}$"* for class control, which could be interpreted as *"if there is a peak at m/z vale 5906, then it is healthy, but if there is no peak at this m/z value, we can not conclude anything"*. Considering *"$Peak_{2022}$ or $Peak_{8867}$"*, there are four possibilities, (1) there are peaks at both m/z values 2022 and 8867, (2) there is only peak at m/z value 2022, (3) there is peak only at m/z value 8867,(4) there is no peak at neither m/z value 2022, nor m/z value 8867. If any of the cases (1)–(3) happens then that sample is diseased, otherwise it is healthy. So, if there is peak at either m/z values 2022 or 8867, shows the occurrence of the target phenotype.

Case Item-sets	Support	Control Item-sets	Support
$Peak_{4442}$	1	$\neg Peak_{4442}$	1
$Peak_{4457}$	1	$\neg Peak_{4457}$	1
$Peak_{4495}$	1	$\neg Peak_{4495}$	1
$Peak_{2022}$ or $Peak_{8867}$	1	$\neg Peak_{4429}$	0.87
$Peak_{2022}$ or $Peak_{8936}$	1	$Peak_{5906}$	0.87
$Peak_{2022}$ or $Peak_{8988}$	1		

6   Conclusion

One of the main challenges in analysing MS-data is to identify the most relevant features from the data in hand and provide an easily understandable and interpretable classifier model to the users like medical doctors. In this study, we are also interested in identifying the inter-dependency between the peptides. In order to find these inter-dependencies, we applied an adaptation of TitanicOR, an AR mining algorithm, to extract the most relevant features for the classification problem and their inter-dependencies. The model driven by this approach is also easily interpretable for the users.

The output from the AR mining algorithm was given to an AR based classification algorithm to do the classification of the new samples.

This approach was able to provide an easily interpretable model to the users which also includes information which was discarded by other greedy ML algorithms which tend to find the minimum number of features.

We also applied six other algorithms, including Binda, SVM, Lasso, Rpart, Jrip and NMF on two different groups of data-sets described in 4. The classification accuracy of our method is 100% for the simulated data-sets and 90% for the real data-set, which is comparable with other methods.

List of abbreviations
MALDI-TOF: Matrix-Assisted Laser Desorption Ionization Time-Of-Flight; ML - Machine Learning; MS - Mass Spectrometry; SVM - Support Vector Machine; TP - True Positive; TN - True Negative; FP - False Positive; FN - False Negative; AR - Association Rule; EP - Emerging Pattern;

7   Declarations

Ethics approval and consent to participate. The ethics committees of the Medical Faculties of the Universities of Leipzig and Heidelberg approved the use of the samples. All subjects gave written informed consent at both centres to participate in the study.

Availability of data and materials. The method source-code can be downloaded from our homepage: http://software.medicalbioinformatics.de. The used data will be made available through request from the authors.

Competing interests. The authors declare that they have no competing interests.

Funding. TC was supported the DFG Research Center Matheon *Mathematics for key technologies*, Berlin. FJ and TC are supported by the German Ministry of Research and Education (BMBF) project Grant 3FO18501 (Forschungscampus MODAL).

Authors' Contributions. Conceived and designed the experiments: FJ, TC. Performed the experiments: FJ. All authors contributed to writing the paper. All authors read and approved the final manuscript.

References

1. Vapnik, V.: Pattern recognition using generalized portrait method. Autom. Remote Control **24**, 774–780 (1963)
2. Fan, R.E., Chang, K.W., Hsieh, C.J., Wang, X.R., Lin, C.J.: Liblinear: A library for large linear classification. J. Mach. Learn. Res. **9**, 1871–1874 (2008)
3. Helleputte, T.: LiblineaR: Linear Predictive Models Based on the LIBLINEAR C/C++ Library. R package version 2.10-8 (2017)
4. Simon, N., Friedman, J., Hastie, T., Tibshirani, R.: Regularization paths for cox's proportional hazards model via coordinate descent. J. Stat. Softw. **39**(5), 1–13 (2011)
5. Friedman, J., Hastie, T., Tibshirani, R.: Regularization paths for generalized linear models via coordinate descent. J. Stat. Softw. **33**(1), 1–22 (2010)
6. Therneau, T., Beth Atkinson, B.R.: Recursive Partitioning and Regression Trees. R package version 4.1-10 (2015)
7. Kuhn, M.: Classification and Regression Training. R package version 6.0-73 (2016)
8. Vimieiro, R., Moscato, P.: Mining disjunctive minimal generators with titanicor. Expert Syst. Appl. **39**(9), 8228–8238 (2012)
9. Gibb, S., Strimmer, K.: Multi-Class Discriminant Analysis using Binary Predictors. R package version 1.0.3 (2015)
10. Holzinger, A.: Interactive machine learning for health informatics: when do we need the human-in-the-loop? Brain Inf. **3**(2), 119–131 (2016)
11. Holzinger, A., Plass, M., Holzinger, K., Crisan, G.C., Pintea, C.M., Palade, V.: A glass-box interactive machine learning approach for solving np-hard problems with the human-in-the-loop. arXiv preprint (2017). arXiv:1708.01104
12. Bakin, S., et al.: Adaptive regression and model selection in data mining problems. Ph.D. thesis, The Australian National University (1999)
13. Lawton, W.H., Sylvestre, E.A.: Self modeling curve resolution. Technometrics **13**(3), 617–633 (1971)
14. Loekito, E., Bailey, J.: Fast mining of high dimensional expressive contrast patterns using zero-suppressed binary decision diagrams. In: Proceedings of the 12th ACM SIGKDD International Conference on Knowledge Discovery and Data Mining, pp. 307–316. ACM (2006)
15. Vimieiro, R., Moscato, P.: A new method for mining disjunctive emerging patterns in high-dimensional datasets using hypergraphs. Inf. Syst. **40**, 1–10 (2014)
16. Vimieiro, R.: Mining disjunctive patterns in biomedical data sets. Ph.D. thesis, University of Newcastle, Faculty of Engineering & Built Environment, School of Electrical Engineering and Computer Science (2012)
17. Zhao, L., Zaki, M.J., Ramakrishnan, N.: Blosom: a framework for mining arbitrary boolean expressions. In: Proceedings of the 12th ACM SIGKDD International Conference on Knowledge Discovery and Data Mining, pp. 827–832. ACM (2006)
18. Liu, Q., Sung, A.H., Qiao, M., Chen, Z., Yang, J.Y., Yang, M.Q., Huang, X., Deng, Y.: Comparison of feature selection and classification for maldi-ms data. BMC Genom. **10**(1), S3 (2009)

19. Swan, A.L., Mobasheri, A., Allaway, D., Liddell, S., Bacardit, J.: Application of machine learning to proteomics data: classification and biomarker identification in postgenomics biology. Omics: J. Integr. Biol. **17**(12), 595–610 (2013)
20. Agrawal, R., Imieliński, T., Swami, A.: Mining association rules between sets of items in large databases. In: ACM Sigmod Record, vol. 22, pp. 207–216. ACM (1993)
21. Varadan, V., Anastassiou, D.: Inference of disease-related molecular logic from systems-based microarray analysis. PLoS Comput. Biol. **2**(6), e68 (2006)
22. Sahoo, D., Dill, D.L., Gentles, A.J., Tibshirani, R., Plevritis, S.K.: Boolean implication networks derived from large scale, whole genome microarray datasets. Genome Biol. **9**(10), R157 (2008)
23. Li, J., Li, H., Wong, L., Pei, J., Dong, G.: Minimum description length principle: Generators are preferable to closed patterns. AAA I, 409–414 (2006)
24. Gibb, S., Strimmer, K.: MALDIquant: a versatile R package for the analysis of mass spectrometry data. Bioinformatics **28**(17), 2270–2271 (2012)
25. Savitzky, A., Golay, M.J.: Smoothing and differentiation of data by simplified least squares procedures. Anal. Chem. **36**(8), 1627–1639 (1964)
26. He, Q.P., Wang, J., Mobley, J.A., Richman, J., Grizzle, W.E.: Self-calibrated warping for mass spectra alignment. Cancer Inf. **10**, 65 (2011)
27. Fayyad, U., Irani, K.: Multi-interval discretization of continuous valued attributes for classification learning. In: Proceedings of the 13th International Joint Conference on Artificial Intelligence, pp. 1022–1029 (1993)
28. Kim, H.: Data preprocessing, discretization for classification. R package version 1.0-1 (2010)
29. Rissanen, J.: Modeling by shortest data description. Automatica **14**(5), 465–471 (1978)
30. Stumme, G., Taouil, R., Bastide, Y., Pasquier, N., Lakhal, L.: Computing iceberg concept lattices with titanic. Data Knowl. Eng. **42**(2), 189–222 (2002)
31. Agrawal, R., Srikant, R., et al.: Fast algorithms for mining association rules. In: Proceedings 20th International Conference Very Large Data Bases, VLDB, vol. 1215, pp. 487–499 (1994)
32. Li, J.: Prediction by collective likelihood from emerging patterns, US Patent Ap. 10/524,606, 22 August 2002
33. Dong, G., Li, J.: Efficient mining of emerging patterns: discovering trends and differences. In: Proceedings of the Fifth ACM SIGKDD International Conference on Knowledge Discovery and Data Mining, pp. 43–52. ACM (1999)
34. Fiedler, G.M., Leichtle, A.B., Kase, J., Baumann, S., Ceglarek, U., Felix, K., Conrad, T., Witzigmann, H., Weimann, A., Schütte, C., et al.: Serum peptidome profiling revealed platelet factor 4 as a potential discriminating peptide associated with pancreatic cancer. Clin. Cancer Res. **15**(11), 3812–3819 (2009)
35. Conrad, T.O., Genzel, M., Cvetkovic, N., Wulkow, N., Leichtle, A., Vybiral, J., Kutyniok, G., Schütte, C.: Sparse proteomics analysis-a compressed sensing-based approach for feature selection and classification of high-dimensional proteomics mass spectrometry data. BMC Bioinf. **18**(1), 160 (2017)

Probabilistic Logic Programming in Action

Arnaud Nguembang Fadja[1(✉)] and Fabrizio Riguzzi[2]

[1] Dipartimento di Ingegneria, University of Ferrara,
Via Saragat 1, 44122 Ferrara, Italy
`arnaud.nguembangfadja@unife.it`
[2] Dipartimento di Matematica e Informatica, University of Ferrara,
Via Saragat 1, 44122 Ferrara, Italy
`fabrizio.riguzzi@unife.it`

Abstract. Probabilistic Programming (PP) has recently emerged as an effective approach for building complex probabilistic models. Until recently PP was mostly focused on functional programming while now Probabilistic Logic Programming (PLP) forms a significant subfield. In this paper we aim at presenting a quick overview of the features of current languages and systems for PLP. We first present the basic semantics for probabilistic logic programs and then consider extensions for dealing with infinite structures and continuous random variables. To show the modeling features of PLP in action, we present several examples: a simple generator of random 2D tile maps, an encoding of Markov Logic Networks, the truel game, the coupon collector problem, the one-dimensional random walk, latent Dirichlet allocation and the Indian GPA problem. These examples show the maturity of PLP.

Keywords: Probabilistic Logic Programming · Probabilistic logical inference · Hybrid program

1 Introduction

Probabilistic Logic Programming (PLP) [11] models domains characterized by complex and uncertain relationships among domain entities by combining probability theory with logic programming. The field started in the early nineties with the seminal work of Dantsin [8], Poole [33] and Sato [49] and is now well established, with a dedicated annual workshop since 2014.

PLP has been applied successfully to many problems such as concept relatedness in biological networks [12], Mendel's genetic inheritance [50], natural language processing [44,51], link prediction in social networks [24], entity resolution [37] and model checking [15].

PLP is a type of Probabilistic Programming (PP) [30], a collection of techniques that have recently emerged as an effective approach for building complex probabilistic models. Until recently PP was mostly focused on functional programming while now PLP forms a significant subfield.

Various approaches have been proposed for representing probabilistic information in Logic Programming [10,27]. The distribution semantics [49] is one of

A. Holzinger et al. (Eds.): Integrative Machine Learning, LNAI 10344, pp. 89–116, 2017.
https://doi.org/10.1007/978-3-319-69775-8_5

the most used and underlies many languages, such as Independent Choice Logic [32], PRISM [49], Logic Programs with Annotated Disjunctions (LPADs) [56] and ProbLog [12]. While these languages differ syntactically, they have the same expressive power, as there are linear transformations among them [55].

In this paper we aim at giving an overview of the current status of PLP. We start by first presenting the semantics for programs without function symbols and then discussing the extensions of this semantics for programs including function symbols, that may have infinite computation branches, and for programs including continuous random variables, a recent proposal that brought PLP closer to functional PP approaches, where continuous random variables have been a basic feature for some time now.

We also discuss approaches for inference, starting from exact inference by knowledge compilation and going to techniques for dealing with infinite computation branches and continuous random variables. Then we illustrate approximate inference approaches based on Monte Carlo methods that can overcome some of the limit of exact inference both in terms of computation time and allowed language features, permitting inference on a less restricted class of programs.

To show the modeling features of PLP in action, we present several examples. We start from a simple generator of random 2D tile maps for video games and from an approach for encoding Markov Logic Networks, a popular Statistical Relation Artificial Intelligence formalism. In both examples all the variables are discrete and no infinite computation path exists. We then consider three domains with possibly infinite computations: the truel problem from game theory, the coupon collector problem and the one-dimensional random walk. Finally we discuss two examples where some of the variables are continuous: latent Dirichlet allocation and the Indian GPA problem. For each example we provide the code for the `cplint` system [1] together with links to the web application `cplint` on SWISH http://cplint.ml.unife.it [39,47] where the examples can be run online, without the need to install the system locally.

PLP has found important applications in the field of Machine Learning and Knowledge Extraction [17]: it is used as the language for representing input data and output models in a variety of Machine Learning systems [14,35,40,43] that have already achieved a significant number of successful applications. The web application `cplint` on SWISH also includes the learning systems EMBLEM [3,4], SLIPCOVER [5] and LEMUR [13]. For lack of space we do not discuss learning in this paper.

2   Syntax and Semantics

We consider first the discrete version of probabilistic logic programming languages. In this version, each logical atom is a Boolean random variable that can assume values true or false. The facts and rules of the program specify the dependences among the truth values of atoms and the main inference task is to compute the probability that a ground query is true, often conditioned on the truth of a conjunction of ground goals, the evidence. All the languages following the distribution semantics allow the specification of alternatives either for facts and/or for clauses. We present here the syntax of LPADs [56] for its generality.

An LPAD is a finite set of annotated disjunctive clauses of the form

$$h_{i1} : \Pi_{i1}; \ldots; h_{in_i} : \Pi_{in_i} :- b_{i1}, \ldots, b_{im_i}.$$

where $b_{i1}, \ldots, b_{im_i}$ are literals, $h_{i1}, \ldots h_{in_i}$ are atoms and $\Pi_{i1}, \ldots, \Pi_{in_i}$ are real numbers in the interval $[0, 1]$ such that $\sum_{k=1}^{n_i} \Pi_{ik} \leq 1$. This clause can be interpreted as "if $b_{i1}, \ldots, b_{im_i}$ is true, then h_{i1} is true with probability Π_{i1} or ... or h_{in_i} is true with probability Π_{in_i}." $b_{i1}, \ldots, b_{im_i}$ is the body of the clause and we indicate it with $body(C)$ if the clause is C. If $n_i = 1$ and $\Pi_{i1} = 1$ the clause is non-disjunctive and so not probabilistic. If $\sum_{k=1}^{n_i} \Pi_{ik} < 1$, there is an implicit annotated atom $null : (1 - \sum_{k=1}^{n_i} \Pi_{ik})$ that does not appear in the body of any clauses of the program.

Given an LPAD P, the grounding $ground(P)$ is obtained by replacing variables with all possible logic terms. If P does not contain function symbols, the set of possible terms is equal to the set of all constants appearing in P and is finite so $ground(P)$ is finite as well.

$ground(P)$ is still an LPAD from which, by selecting a head atom for each ground clause, we can obtain a normal logic program, called "world". In the distribution semantics, the choices of head atoms for different clauses are independent, so we can assign a probability to a world by multiplying the probabilities of all the head atoms chosen to form the world. In this way we get a probability distribution over worlds from which we can define a probability distribution over the truth values of a ground atom: the probability of an atom q being true is the sum of the probabilities of the worlds where q is true in the well-founded model [54] of the world.

The well-founded model [54] in general is three valued, so a query that is not true in the model is either undefined or false. However, we consider atoms as Boolean random variables so we do not want to deal with the undefined truth value. How to manage uncertainty by combining nonmonotonic reasoning and probability theory is still an open problem, see [7] for a discussion of the issues. So we require each world to have a two-valued well-founded model and therefore q can only be true or false in a world.

Formally, each grounding of a clause $C_i \theta_j$ corresponds to a random variable X_{ij} with as many values as the number of head atoms of C_i. The random variables X_{ij} are independent of each other.

An *atomic choice* [31] is a triple (C_i, θ_j, k) where $C_i \in P$, θ_j is a substitution that grounds C_i and $k \in \{1, \ldots, n_i\}$ identifies one of the head atoms. In practice $C_i \theta_j$ corresponds to an assignment $X_{ij} = k$. A set of atomic choices κ is *consistent* if only one head is selected for the same ground clause. A consistent set κ of atomic choices is called a *composite choice*. We can assign a probability to κ as the random variables are independent: $P(\kappa) = \prod_{(C_i, \theta_j, k) \in \kappa} \Pi_{ik}$.

A *selection* σ is a composite choice that, for each clause $C_i \theta_j$ in $ground(P)$, contains an atomic choice (C_i, θ_j, k). A selection σ identifies a normal logic program l_σ defined as $l_\sigma = \{(h_{ik} \leftarrow body(C_i))\theta_j | (C_i, \theta_j, k) \in \sigma\}$. l_σ is called an *instance, possible world* or simply *world* of P. Since selections are composite choices, we can assign a probability to instances: $P(l_\sigma) = P(\sigma) = \prod_{(C_i, \theta_j, k) \in \sigma} \Pi_{ik}$.

We write $l_\sigma \models q$ to mean that the query q (a ground atom) is true in the well-founded model of the program l_σ. The probability of a query q given a world l_σ can be now defined as $P(q|l_\sigma) = 1$ if $l_\sigma \models q$ and 0 otherwise. Let $P(L_P)$ be the distribution over worlds. The probability of a query q is given by

$$P(q) = \sum_{l_\sigma \in L_P} P(q, l_\sigma) = \sum_{l_\sigma \in L_P} P(q|l_\sigma)P(l_\sigma) = \sum_{l_\sigma \in L_P : l_\sigma \models q} P(l_\sigma) \qquad (1)$$

Example 1 (From [5]). The following LPAD P encodes geological knowledge on the Stromboli Italian island:
$C_1 = eruption : 0.6 \; ; \; earthquake : 0.3 :- \; sudden\_energy\_release,$
    $fault\_rupture(X).$
$C_2 = sudden\_energy\_release : 0.7.$
$C_3 = fault\_rupture(southwest\_northeast).$
$C_4 = fault\_rupture(east\_west).$
The Stromboli island is located at the intersection of two geological faults, one in the southwest-northeast direction, the other in the east-west direction, and contains a active volcano. This program models the possibility that an eruption or an earthquake occurs at Stromboli. If there is a sudden energy release under the island and there is a fault rupture, then there can be an eruption of the volcano on the island with probability 0.6 or an earthquake in the area with probability 0.3 or no event with probability 0.1. The energy release occurs with probability 0.7 while we are sure that ruptures occur in both faults.

Clause C_1 has two groundings, $C_1\theta_1$ with $\theta_1 = \{X/southwest\_northeast\}$ and $C_1\theta_2$ with $\theta_2 = \{X/east\_west\}$, so there are two random variables X_{11} and X_{12}. Clause C_2 has only one grounding $C_2\emptyset$ instead, so there is one random variable X_{21}. X_{11} and X_{12} can take three values since C_1 has three head atoms; similarly X_{21} can take two values since C_2 has two head atoms. P has 18 instances, the query *eruption* is true in 5 of them and its probability is $P(eruption) = 0.6 \cdot 0.6 \cdot 0.7 + 0.6 \cdot 0.3 \cdot 0.7 + 0.6 \cdot 0.1 \cdot 0.7 + 0.3 \cdot 0.6 \cdot 0.7 + 0.1 \cdot 0.6 \cdot 0.7 = 0.588$.

This semantics can be given also a sampling interpretation: the probability of a query q is the fraction of worlds, sampled from the distribution over worlds, where q is true. To sample from the distribution over worlds, you simply randomly select a head atom for each ground clause according to the probabilistic annotations. Note that you don't even need to sample a complete world: if the samples you have taken ensure the truth value of q is determined, you don't need to sample more clauses, as they don't influence q.

To compute the conditional probability $P(q|e)$ of a query q given evidence e, you can use the definition of conditional probability, $P(q|e) = P(q, e)/P(e)$, and compute first the probability of q, e (the sum of probabilities of worlds where both q and e are true) and the probability of e and then divide the two.

If the program P contains function symbols, a more complex definition of the semantics is necessary. In fact $ground(P)$ is infinite and a world would be obtained by making an infinite number of choices so its probability would be 0, as it is a product of infinite numbers all bounded away from 1 from below. In this case we have to work with sets of worlds and use Kolmogorov's definition of probability space. It turns out that the probability of the query is the sum of a convergent series [38].

Up to now we have considered only discrete random variables and discrete probability distributions. How can we consider continuous random variables and probability density functions, for example real variables following a Gaussian distribution? cplint [1] and Distributional Clauses (DC) [28] allow the description of continuous random variables in so called *hybrid programs*.

cplint allows the specification of density functions over arguments of atoms in the head of rules. For example, in

```
g(X,Y): gaussian(Y,0,1):- object(X).
```

X takes terms while Y takes real numbers as values. The clause states that, for each X such that object(X) is true, the values of Y such that g(X,Y) is true, follow a Gaussian distribution with mean 0 and variance 1. You can think of an atom such as g(a,Y) as an encoding of a continuous random variable associated to term g(a). In DC you can express the same density as

```
g(X)~gaussian(0,1):= object(X).
```

where := indicates implication and continuous random variables are represented as terms that denote a value from a continuous domain. It is possible to translate DC into programs for cplint and in fact cplint allows also the DC syntax, automatically translating DC into its own syntax.

A semantics for hybrid programs was given independently in [16,20,28]. In [28] the semantics of Hybrid Probabilistic Logic Programs (HPLP) is defined by means of a stochastic generalization STp of the Tp operator that applies the sampling interpretation of the distribution semantics to continuous variables: STp is applied to interpretations that contain ground atoms (as in standard logic programming) and terms of the form $t = v$ where t is a term indicating a continuous random variable and v is a real number. If the body of a clause is true in an interpretation I, $STp(I)$ will contain a sample from the head.

The authors of [20] define a probability space for N continuous random variables by considering the Borel σ-algebra over $\mathbb{R}^N$ and fixing a Lebesgue measure on this set as the probability measure. The probability space is lifted to cover the entire program using the least model semantics of constraint logic programs.

If an atom encodes a continuous random variable (such as g(X,Y) above), asking for the probability that a ground instantiation, such as g(a,0.3), is true is not meaningful, as the probability that a continuous random variables takes a specific value is always 0. In this case you want to compute the probability that the random variable falls in an interval or you want to know its density, possibly after having observed some evidence. If the evidence is on an atom defining another continuous random variable, the definition of conditional probability cannot be applied, as the probability of the evidence would be 0 and so the fraction would be undefined. This problem is tackled in [28] by providing a definition using limits.

3 Inference

Computing all the worlds is impractical because their number is exponential in the number of ground probabilistic clauses when there are no function symbols and impossible otherwise, because with function symbols the number of worlds is uncountably infinite. Alternative approaches for inference have been considered that can be grouped in exact and approximate ones [1].

For exact inference from discrete program without function symbols a successful approach finds explanations for the query q [12], where an explanation is a set of clause choices that are sufficient for entailing the query. Once all explanations for the query are found, they are encoded as a Boolean formula in DNF and the problem is reduced to that of computing the probability that the formula is true given the probabilities of being true of all the (mutually independent) random variables. This problem is called *disjoint-sum* as it can be solved by finding a DNF where all the disjuncts are mutually exclusive. Its complexity is #P [53] so the problem is highly difficult and intractable in general. In practice, problems of significant size can be tackled using *knowledge compilation* [9], i.e. converting the DNF into a language from which the computation of the probability is polynomial [12,45], such as Binary Decision Diagrams.

Formally, a composite choice κ is an *explanation* for a query q if q is entailed by every instance consistent with κ, where an instance l_σ is consistent with κ iff $\kappa \subseteq \sigma$. Let λ_κ be the set of worlds consistent with κ. In particular, algorithms find a covering set of explanations for the query, where a set of composite choices K is *covering* with respect to q if every program l_σ in which q is entailed is in λ_K, where $\lambda_K = \sum_{\kappa \in K} \lambda_\kappa$. The problem of computing the probability of a query q can thus be reduced to computing the probability of the Boolean function

$$f_q(\mathbf{X}) = \bigvee_{\kappa \in E(q)} \bigwedge_{(C_i, \theta_j, k) \in \kappa} X_{ij} = k \qquad (2)$$

where $E(q)$ is a covering set of explanations for q.

Example 2 (Example 1 cont.). The query *eruption* has the covering set of explanations $E(eruption) = \{\kappa_1, \kappa_2\}$ where:

$$\kappa_1 = \{(C_1, \{X/southwest\_northeast\}, 1), (C_2, \{\}, 1)\}$$
$$\kappa_2 = \{(C_1, \{X/east\_west\}, 1), (C_2, \{\}, 1)\}$$

Each atomic choice (C_i, θ_j, k) is represented by the propositional equation $X_{ij} = k$:

$$\begin{aligned}
(C_1, \{X/southwest\_northeast\}, 1) &\rightarrow X_{11} = 1 \\
(C_1, \{X/east\_west\}, 1) &\rightarrow X_{12} = 1 \\
(C_2, \{\}, 1) &\rightarrow X_{21} = 1
\end{aligned}$$

The resulting Boolean function $f_{eruption}(\mathbf{X})$ returns 1 if the values of the variables correspond to an explanation for the goal. Equations for a single explanation are conjoined and the conjunctions for the different explanations are disjoined. The set of explanations $E(eruption)$ can thus be encoded with the function:

$$f_{eruption}(\mathbf{X}) = (X_{11} = 1 \wedge X_{21} = 1) \vee (X_{12} = 1 \wedge X_{21} = 1) \qquad (3)$$

Examples of systems that perform inference using this approach are ProbLog [23] and PITA [45,46]. Recent approaches for exact inference try to achieve speedups by reasoning at a lifted level [2,41].

When a discrete program contains function symbols, the number of explanations may be infinite and the probability of the query may be the sum of a convergent series. In this case the inference algorithm has to recognize the presence of an infinite number of explanations and identify the terms of the series. In [48,52] the authors extended PRISM by considering programs under the *generative exclusiveness condition*: at any choice point in any execution path of the top-goal, the choice is done according to a value sampled from a PRISM probabilistic switch. The generative exclusiveness condition implies that every disjunction is exclusive and originates from a probabilistic choice made by some switch.

In this case, a *cyclic explanation graph* can be computed that encodes the dependence of atoms on probabilistic switches. From this a system of equations can be obtained defining the probability of ground atoms. The authors of [48,52] show that by first assigning all atoms probability 0 and repeatedly applying the equations to compute updated values results in a process that converges to a solution of the system of equations. For some program, such as those computing the probability of prefixes of strings from Probabilistic Context Free Grammars, the system is linear, so solving it is even simpler. In general, this provides an approach for performing inference when the number of explanations is infinite but under the generative exclusiveness condition.

In [15] the authors present the algorithm PIP (for Probabilistic Inference Plus), that is able to perform inference even when explanations are not necessarily mutually exclusive and the number of explanations is infinite. They require the programs to be *temporally well-formed*, i.e., that one of the arguments of predicates can be interpreted as a time that grows from head to body. In this case the explanations for an atom can be represented succinctly by Definite Clause Grammars (DCGs). Such DCGs are called *explanation generators* and are used to build Factored Explanation Diagrams (FED) that have a structure that closely follows that of Binary Decision Diagrams. FEDs can be used to obtain a system of polynomial equations that is monotonic and thus convergent as in [48,52]. So, even when the system is non linear, a least solution can be computed to within an arbitrary approximation bound by an iterative procedure.

For approximate inference one of the most used approach consists in Monte Carlo sampling, following the sampling interpretation of the semantics given above. Monte Carlo backward reasoning has been implemented in ProbLog [23] and MCINTYRE [36] and found to give good performance in terms of quality of the solutions and of running time. Monte Carlo sampling is attractive for the simplicity of its implementation and because you can improve the estimate as more time is available. Moreover, Monte Carlo can be used also for programs with function symbols, in which goals may have infinite explanations and exact inference may loop. In fact, taking a sample of a query corresponds naturally to an explanation and the probability of a derivation is the same as the probability of the corresponding explanation. The risk is that of incurring in an infinite explanation. But infinite explanations have probability 0 so the probability that the computation goes down such a path and does not terminate is 0 as well.

Monte Carlo inference provides also smart algorithms for computing conditional probabilities: rejection sampling or Metropolis-Hastings Markov Chain Monte Carlo (MCMC). In rejection sampling [57], you first query the evidence and, if the query is successful, query the goal in the same sample, otherwise the sample is discarded.

In Metropolis-Hastings MCMC [26], a Markov chain is built by taking an initial sample and by generating successor samples. The initial sample is built by randomly sampling choices so that the evidence is true. A successor sample is obtained by deleting a fixed number of sampled probabilistic choices. Then the evidence is queried again by sampling starting with the undeleted choices. If the query succeeds, the goal is then also queried by sampling. The goal sample is accepted with a probability of $\min\{1, \frac{N_0}{N_1}\}$ where N_0 is the number of choices sampled in the previous sample and N_1 is the number of choices sampled in the current sample. The number of successes of the query is increased by 1 if the query succeeded in the last accepted sample. The final probability is given by the number of successes over the total number of samples.

When you have evidence on ground atoms that have continuous values as arguments, you can still use Monte Carlo sampling. You cannot use rejection sampling or Metropolis-Hastings, as the probability of the evidence is 0, but you can use likelihood weighting [28] to obtain weighted samples of continuous arguments of a goal. For each sample to be taken, likelihood weighting samples the query and then assigns a weight to the sample on the basis of evidence. The weight is computed by deriving the evidence backward in the same sample of the query starting with a weight of one: each time a choice should be taken or a continuous variable sampled, if the choice/variable has already been taken, the current weight is multiplied by probability of the choice/by the density value of the continuous value.

If likelihood weighting is used to find the posterior density of a continuous random variable, we obtain a set of weighted samples for the variables whose weight can be interpreted as a relative frequency. The set of samples without the weight, instead, can be interpreted as the prior density of the variable. These two sets of samples can be used to plot the density before and after observing the evidence.

You can sample arguments of queries also for discrete goals: in this case you get a discrete distribution over the values of one or more arguments of a goal. If the query predicate is determinate in each world, i.e., given values for input arguments there is a single value for output arguments that make the query true, for each sample you get a single value. Moreover, if clauses sharing an atom in the head are mutually exclusive, i.e., in each world the body of at most one clause is true, then the query defines a probability distribution over output arguments. In this way we can simulate those languages such as PRISM and Stochastic Logic Programs [25] that define probability distributions over arguments rather than probability distributions over truth values of ground atoms.

4 Examples

Here we present some examples of PLP in practice. In the first two examples, tile map generation and Markov Logic Networks encoding, all the variables are discrete and no infinite computation path exists.

The next three problems have infinite computation paths: the truel game, the coupon collector problem and the one-dimensional random walk.

The last two examples include continuous variables: latent Dirichlet allocation and the Indian GPA problem.

4.1 Tile Map Generation

PP and PLP can be used to generate random complex structures. For example, we can write programs for randomly generating maps of video games. We are given a fixed set of tiles that we want to combine to obtain a 2D map that is random but satisfies some soft constraints on the placement of tiles.

Suppose we want to draw a 10×10 map with a tendency to have a lake in the center. The tiles are randomly placed such that, in the central area, water is more probable. The problem can be modeled with the following example[1], where map(H,W,M) instantiates M to a map of height H and width W:

```
map(H,W,M):-
  tiles(Tiles),
  length(Rows,H),
  M=..[map,Tiles|Rows],
  foldl(select(H,W),Rows,1,_).

select(H,W,Row,N0,N):-
  length(RowL,W),
  N is N0+1,
  Row=..[row|RowL],
  foldl(pick_row(H,W,N0),RowL,1,_).

pick_row(H,W,N,T,M0,M):-
  M is M0+1,
  pick_tile(N,M0,H,W,T).
```

where foldl/4 is a SWI-Prolog [58] library predicate that implements the foldl meta primitive from functional programming. pick_tile(Y,X,H,W,T) returns in T a tile for position (X,Y) of a map of size W*H. The center tile is water:

---

[1] http://cplint.ml.unife.it/example/inference/tile_map.swinb.

```
pick_tile(HC,WC,H,W,water):-
  HC is H//2,
  WC is W//2,!.
```

In the central area water is more probable:

```
pick_tile(Y,X,H,W,T):
  discrete(T,[grass:0.05,water:0.9,tree:0.025,rock:0.025]):-
  central_area(Y,X,H,W),!
```

central_area(Y, X, H, W) is true if (X, Y) is adjacent to the center of the W*H map (definition omitted for brevity). In the other places, tiles are chosen at random with distribution [grass:0.5,water:0.3,tree:0.1,rock:0.1]:

```
pick_tile(_,_,_,_,T):discrete(T,[grass:0.5,water:0.3,tree:0.1,rock:0.1]).
```

We can generate a map by taking a sample of the query map(10,10,M) and collecting the value of M. For example, the map of Fig. 1 can be obtained[2].

Fig. 1. A random tile map.

---

[2] Tiles from https://github.com/silveira/openpixels.

4.2   Markov Logic Networks

Markov Networks (MN) and Markov Logic Networks (MLN) [34] can be encoded with PLP. The encoding is based on the observation that a MN factor can be represented with a Bayesian Network (BN) with an extra node that is always observed. Since PLP programs under the distribution semantics can encode BN [56], we can encode MLN. An example of an MLN clause is

$$1.5 \; Intelligent(x) \Rightarrow GoodMarks(x)$$

where 1.5 is the weight of the clause.

For a single constant $anna$, this clause originates an edge between the Boolean nodes for $Intelligent(anna)$ and $GoodMarks(anna)$. This means that the two variables cannot be d-separated in any way. This dependence can be modeled with BN by adding and extra Boolean node, $Clause(anna)$, that is a child of $Intelligent\,(anna)$ and $GoodMarks(anna)$ and is observed. In this way, $Intelligent(anna)$ and $GoodMarks(anna)$ are not d-separated in the BN no matter what other nodes the BN contains.

In general, for a domain with Herbrand base X and an MLN ground clause C mentioning atom variables X', the equivalent BN should contain a Boolean node C with X' as parents. All the query of the form $P(a|b)$ should then be posed to the BN as $P(a|b, C = true)$. The problem is now how to assign values to the conditional probability (CPT) of C given X' so that the joint distribution of X in the BN is the same as that of the MLN.

A ground MLN formulae of the form $\alpha\ C$ contributes to the probabilities of the worlds with a factor e^α for the worlds where the clause is true and 1 for the worlds where the clause is false. If we use c to indicate C = true, the joint probability of a state of the world x can then be computed as

$$P(x|c) = \frac{P(x,c)}{P(c)} \propto P(x,c)$$

i.e. $P(x|c)$ is proportional to $P(x,c)$, because the denominator does not depend on x and is thus a normalizing constant.

$P(x,c)$ can be written as

$$P(x,c) = P(c|x)P(x) = P(c|x')P(x)$$

where x' is the state of the parents of C, so

$$P(x|c) \propto P(c|x')P(x)$$

To model the MLN formula we just have to ensure that $P(c|x')$ is proportional to e^α when x' makes C true and to 1 when x' makes C false. We cannot use e^α directly in the CPT for C because it can be larger than 1 but we can use the values $e^\alpha/(1+e^\alpha)$ and $1/(1+e^\alpha)$ that are proportional to e^α and 1 and are surely less than 1.

For an MLN containing the example formula above, the probability of a world would be represented by $P(i, g|c)$ where i and g are values for $Intelligent(anna)$ and $GoodMarks(anna)$ and c is $Clause(anna) = true$. The CPT will have the values $e^{1.5}/(1 + e^{1.5})$ for $Clause(anna)$ being true given that the parents' values make the clause true and $1/(1 + e^{1.5})$ is the probability of $Clause(anna)$ being true given that the parents' values make the clause false.

In order to model MLN formulas with LPADs, we can add an extra atom $clause_i(X)$ for each formula $F_i = \alpha_i \, C_i$ where X is the vector of variables appearing in C_i. Then, when we query for the probability of query q given evidence e, we have to ask for the probability of q given $e \wedge ce$ where ce is the conjunction of all the groundings of $clause_i(X)$ for all values of i. Then, clause C_i should be transformed into a Disjunctive Normal Form (DNF) formula $C_{i1} \vee \ldots \vee C_{in_i}$ where the disjuncts are mutually exclusive and the LPAD should contain the clauses

$$clause_i(X) : e^\alpha/(1 + e^\alpha) \leftarrow C_{ij}$$

for all j. Similarly, $\neg C_i$ should be transformed into a disjoint sum $D_{i1} \vee \ldots \vee D_{im_i}$ and the LPAD should contain the clauses

$$clause_i(X) : 1/(1 + e^\alpha) \leftarrow D_{il}$$

for all l.

Alternatively, if α is negative, e^α will be smaller than 1 and we can use the two probability values e^α and 1 with the clauses

$$clause_i(X) : e^\alpha \leftarrow C_{ij}$$
$$\ldots$$
$$clause_i(X) \leftarrow D_{il}$$

This solution has the advantage that some clauses are certain, reducing the number of random variables. If α is positive in formula $\alpha \, C$, we can consider $-\alpha \, \neg C$.

MLN formulas can also be added to a regular probabilistic logic program. In this case their effect is equivalent to a soft form of evidence, where certain worlds are weighted more than others. This is the same as soft evidence in Figaro [30]. MLN hard constraints, i.e., formulas with an infinite weight, can instead be used to rule out completely certain worlds, those violating the constraint. For example, given hard constraint C equivalent to the sum $C_{i1} \vee \ldots \vee C_{in_i}$, the LPAD should contain the clauses

$$clause_i(X) \leftarrow C_{ij}$$

for all j, and the evidence should contain $clause_i(x)$ for all groundings x of X. In this way, the worlds that violate C are ruled out. Let see an example[3] where we translate the MLN .

```
1.5 Intelligent(x) => GoodMarks(x)
1.1 Friends(x, y) => (Intelligent(x) <=> Intelligent(y))
```

[3] http://cplint.ml.unife.it/example/inference/mln.swinb.

The first MLN formula is translated into

```
clause1(X): 0.8175744762:- \+intelligent(X).
clause1(X): 0.1824255238:- intelligent(X), \+good_marks(X).
clause1(X): 0.8175744762:- intelligent(X), good_marks(X).
```

where $0.8175744762 = e^{1.5}/(1 + e^{1.5})$ and $0.1824255238 = 1/(1 + e^{1.5})$.
    The MLN formula

```
1.1 Friends(x, y) => (Intelligent(x) <=> Intelligent(y))
```

is translated into the clauses

```
clause2(X,Y): 0.7502601056:-
  \+friends(X,Y).
clause2(X,Y): 0.7502601056:-
  friends(X,Y), intelligent(X),intelligent(Y).
clause2(X,Y): 0.7502601056:-
  friends(X,Y), \+intelligent(X),\+intelligent(Y).
clause2(X,Y): 0.2497398944:-
  friends(X,Y), intelligent(X),\+intelligent(Y).
clause2(X,Y): 0.2497398944:-
  friends(X,Y), \+intelligent(X),intelligent(Y).
```

where $0.7502601056 = e^{1.1}/(1+e^{1.1})$ and $0.2497398944 = 1/(1+e^{1.1})$. A priori we
have a uniform distribution over student intelligence, good marks and friendship:

```
intelligent(_):0.5.
good_marks(_):0.5.
friends(_,_):0.5.
```

and there are two students:

```
student(anna).
student(bob).
```

The evidence must include the truth of all groundings of the $clause_i$ predicates:

```
evidence_mln:- clause1(anna),clause1(bob),clause2(anna,anna),
    clause2(anna,bob),clause2(bob,anna),clause2(bob,bob).
```

We want to query the probability that Anna gets good marks given that she is
friend with Bob and Bob is intelligent, so we define

```
ev_intelligent_bob_friends_anna_bob:-
    intelligent(bob),friends(anna,bob),evidence_mln.
```

and query for $P(\text{good\_marks}(\text{anna})|\text{ev\_intelligent\_bob\_friends\_anna\_bob})$
obtaining 0.7330 which is higher than the prior probability 0.6069 of Anna
getting good marks, obtained with the query $P(\text{good\_marks}(\text{anna})|\text{evidence\_}$
mln).

4.3   Truel

A truel [22] is a duel among three opponents. There are three truelists, a, b and c, that take turns in shooting with a gun. The firing order is a, b and c. Each truelist can shoot at another truelist or at the sky (deliberate miss). The truelists have these probabilities of hitting the target (if they are not aiming at the sky): 1/3, 2/3 and 1 for a, b and c respectively. The aim for each truelist is to kill all the other truelists. The question is: what should a do to maximize his probability of winning? Aim at b, c or the sky?

Let us see first the strategy for the other truelists and situations. When only two players are left, the best strategy is to shoot at the other player.

When all three players remain, the best strategy for b is to shoot at c, since if c shoots at him he his dead and if c shoots at a, b remains with c which is the best shooter. Similarly, when all three players remain, the best strategy for c is to shoot at b, since in this way he remains with a, the worst shooter.

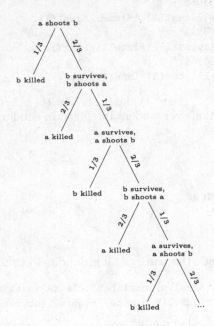

Fig. 2. Probability tree of the truel with opponents a and b.

For a it is more complex. Let us first compute the probability of a to win a duel with a single opponent. When a and c remain, a wins if it shoots c, with probability 1/3. If he misses c, c will surely kill him. When a and b remain, the probability p of a to win can be computed with

$$p = P(\text{a hits b}) + P(\text{a misses b})P(\text{b misses a})p$$
$$p = 1/3 + 2/3 \times 1/3 \times p$$
$$p = 3/7$$

The probability can be also computed by building the probability tree of Fig. 2. The probability that a survives is thus

$$p = 1/3 + 2/3 \cdot 1/3 \cdot 1/3 + 2/3 \cdot 1/3 \cdot 2/3 \cdot 1/3 \cdot 1/3 + \ldots =$$

$$= 1/3 + 2/3^3 + 2^2/3^5 + \ldots = \frac{1}{3} + \sum_{i=0}^{\infty} \frac{2}{3^3} \left(\frac{2}{9}\right)^i = \frac{1}{3} + \frac{\frac{2}{3^3}}{1 - \frac{2}{9}} =$$

$$= \frac{1}{3} + \frac{\frac{2}{3^3}}{\frac{7}{9}} = \frac{1}{3} + \frac{\frac{2}{3}}{7} = \frac{1}{3} + \frac{2}{21} = \frac{9}{21} = \frac{3}{7}$$

When all three players remain, if a shoots at b, b is dead with probability 1/3 but then c will kill a. If b is not dead (probability 2/3), b shoots at c and kills him with probability 2/3. In this case, a is left in a duel with b, with probability of surviving of 3/7. If b doesn't kill c (probability 1/3), c will kill b surely and a is left in a duel with c, with a probability of surviving of 1/3. So overall, if a shoots at b, his probability of winning is

$$2/3 \cdot 2/3 \cdot 3/7 + 2/3 \cdot 1/3 \cdot 1/3 = 4/21 + 2/27 = \frac{36 + 15}{189} = \frac{50}{189} = 0.2645$$

When all three players remain, if a shoots at c, c is dead with probability 1/3. b then shoots at a and a survives with probability 1/3 and a is then in a duel with b and surviving with probability 3/7. If c survives (probability 2/3), b shoots at c and kills him with probability 2/3, so a remains in a duel with b and wins with probability 3/7. If c survives again, he kills b surely and a is left in a duel with c, with probability 1/3 of winning. So overall, if a shoots at c, his probability of winning is

$$1/3 \cdot 1/3 \cdot 3/7 + 2/3 \cdot 2/3 \cdot 3/7 + 2/3 \cdot 1/3 \cdot 1/3 = 1/21 + 4/21 + 2/27 = 59/189 = 0.3122$$

When all three players remain, if a shoots at the sky, b shoots at c and kills him with probability 2/3, with a remaining in a duel with b. If b doesn't kill c, c will surely kill b and a remains in a duel with c. So overall, if a shoots at the sky, his probability of winning is

$$2/3 \cdot 3/7 + 1/3 \cdot 1/3 = 2/7 + 1/9 = 25/63 = 0.3968$$

This problem can be modeled with an LPAD[4]. However, as can be seen from Fig. 2, the number of explanations may be infinite so we have to use an appropriate exact inference algorithm or Monte Carlo inference. We discuss below a program that uses MCINTYRE.

survives_action(A, L0, T, S) is true if A survives truel performing action S with L0 still alive in turn T:

```
survives_action(A,L0,T,S):-
  shoot(A,S,L0,T,L1),
  remaining(L1,A,Rest),
  survives_round(Rest,L1,A,T).
```

---

[4] http://cplint.ml.unife.it/example/inference/truel.pl.

shoot(H,S,L0,T,L) is true when H shoots at S in round T with L0 and L the list of truelists still alive before and after the shot:

```
shoot(H,S,L0,T,L):-
    (S=sky -> L=L0
  ;  (hit(T,H) ->  delete(L0,S,L)
    ;  L=L0
    )
  ).
```

The probabilities of each truelist to hit the chosen target are

```
hit(_,a):1/3.
hit(_,b):2/3.
hit(_,c):1.
```

survives(L,A,T) is true if individual A survives the truel with truelists L at round T:

```
survives([A],A,_):-!.

survives(L,A,T):-
  survives_round(L,L,A,T).
```

survives_round(Rest, L0, A, T) is true if individual A survives the truel at round T with Rest still to shoot and L0 still alive:

```
survives_round([],L,A,T):-
  survives(L,A,s(T)).

survives_round([H|_Rest],L0,A,T):-
  base_best_strategy(H,L0,S),
  shoot(H,S,L0,T,L1),
  remaining(L1,H,Rest1),
  member(A,L1),
  survives_round(Rest1,L1,A,T).
```

These strategies are easy to find:

```
base_best_strategy(b,[b,c],c).
base_best_strategy(c,[b,c],b).
base_best_strategy(a,[a,c],c).
base_best_strategy(c,[a,c],a).
base_best_strategy(a,[a,b],b).
base_best_strategy(b,[a,b],a).
base_best_strategy(b,[a,b,c],c).
base_best_strategy(c,[a,b,c],b).
```

Auxiliary predicate remaining/3 is defined as

```
remaining([A|Rest],A,Rest):-!.

remaining([_|Rest0],A,Rest):-
  remaining(Rest0,A,Rest).
```

We can decide the best strategy for a by asking the queries

```
survives_action(a,[a,b,c],0,b)
survives_action(a,[a,b,c],0,c)
survives_action(a,[a,b,c],0,sky)
```

If we take 1000 samples, possible answers are 0.256, 0.316 and 0.389, showing that a should aim at the sky.

4.4   Coupon Collector Problem

The coupon collector problem is described in [21] as

> Suppose each box of cereal contains one of N different coupons and once a consumer has collected a coupon of each type, he can trade them for a prize. The aim of the problem is determining the average number of cereal boxes the consumer should buy to collect all coupon types, assuming that each coupon type occurs with the same probability in the cereal boxes.

If there are N different coupons, how many boxes, T, do I have to buy to get the prize? This problem can be modeled by a program[5] defining predicate coupons/2 such that coupons(N,T) is true if we need T boxes to get N coupons. We represent the coupons with a term for functor cp/N with the number of coupons as arity. The ith argument of the term is 1 if the ith coupon has been collected and is a variable otherwise. The term thus represents an array:

```
coupons(N,T):-
  length(CP,N),
  CPTerm=..[cp|CP],
  new_coupon(N,CPTerm,0,N,T).
```

If 0 coupons remain to be collected, the collection ends:

```
new_coupon(0,_CP,T,_N,T).
```

If NO coupons remain to be collected, collect one and recurse:

```
new_coupon(NO,CP,TO,N,T):-
  NO>0,
  collect(CP,N,TO,T1),
  N1 is NO-1,
  new_coupon(N1,CP,T1,N,T).
```

collect/4 collects one new coupon and updates the number of boxes bought:

```
collect(CP,N,TO,T):-
  pick_a_box(TO,N,I),
  T1 is TO+1,
  arg(I,CP,CPI),
  (var(CPI)-> CPI=1, T=T1
  ; collect(CP,N,T1,T)
  ).
```

---

[5] http://cplint.ml.unife.it/example/inference/coupon.swinb.

`pick_a_box/3` randomly picks a box, an element from the list $[1 \ldots N]$:

```
pick_a_box(_,N,I):uniform(I,L):- numlist(1, N, L).
```

If there are 5 different coupons, we may ask:

- how many boxes do I have to buy to get the prize?
- what is the distribution of the number of boxes I have to buy to get the prize?
- what is the expected number of boxes I have to buy to get the prize?

To answer the first query, we can take a single sample for the query `coupons(5,T)`: in the sample, the query will succeed as `coupons/2` is a determinate predicate and the result will instantiate T to a specific value. For example, we may get T=15. Note that the maximum number of boxes to buy is unbounded but the case where we have to buy an infinite number of boxes has probability 0, so sampling will surely finish.

To compute the distribution on the number of boxes, we can take a number of samples, say 1000, and plot the number of times a value is obtained as a function of the value. We can do so by dividing the domain of the number of boxes in intervals and counting the number of sampled values that fall in each interval. By doing so we may get the graph in Fig. 3.

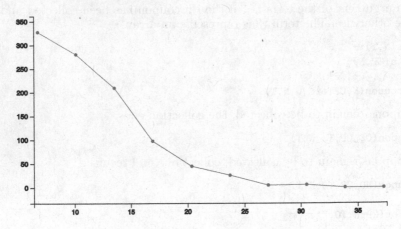

Fig. 3. Distribution of the number of boxes.

To compute the expected number of boxes, we can take a number of samples, say 100, of `coupons(5,T)`. Each sample will instantiate T. By summing all these values and dividing the 100, the number of samples, we can get an estimate of the expectation. For example, we may get a value of 11.47.

We can also plot the dependency of the expected number of boxes from the number of coupons, obtaining Fig. 4. As observed in [21], the number of boxes grows as $O(N \log N)$ where N is the number of coupons. The graph shows the accordance of the two curves.

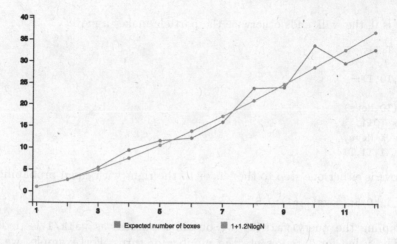

Fig. 4. Expected number of boxes as a function the number of coupons

The coupon collector problem is similar to the sticker collector problem, where you have an album with a space for every different sticker, you can buy stickers in packs and your objective is to complete the album. A program for the coupon collector problem can be applied to solve the sticker collector problem: if you have N different stickers and packs contain P stickers, we can solve the coupon collector problem for N coupons and get the number of boxes B. Then the number of packs you have to buy to complete the collection is $\lceil B/P \rceil$. So we can write:

```
stickers(N,P,T):- coupons(N,T0), T is ceiling(T0/P).
```

If there are 50 different stickers and packs contain 4 stickers, by sampling the query `stickers(50,4,T)` we can get T=47, i.e., we have to buy 47 packs to complete the entire album.

4.5 One-Dimensional Random Walk

We consider the version of the problem described in [21]: a particle starts at position $x = 10$ and moves with equal probability one unit to the left or one unit to the right in each turn. The random walk stops if the particle reaches position $x = 0$.

The walk terminates with probability one [19] but requires, on average, an infinite time, i.e., the expected number of turns is infinite [21].

We can compute the number of turns with the following program[6]. The walk starts at time 0 and $x = 10$:

```
walk(T):- walk(10,0,T).
```

---

[6] http://cplint.ml.unife.it/example/inference/random_walk.swinb.

If x is 0, the walk ends otherwise the particle makes a move:

```
walk(0,T,T).

walk(X,T0,T):-
  X>0,
  move(T0,Move),
  T1 is T0+1,
  X1 is X+Move,
  walk(X1,T1,T).
```

The move is either one step to the left or to the right with equal probability.

```
move(T,1):0.5; move(T,-1):0.5.
```

By sampling the query `walk(T)` we obtain a success as `walk/1` is determinate. The value for `T` represents the number of turns. For example, we may get `T = 3692`.

4.6   Latent Dirichlet Allocation

Text mining [18] aims at extracting knowledge from texts. Latent Dirichlet Allocation (LDA) [6] is a text mining technique which assigns topics from a finite set to words in documents. The model describes a generative process where documents are represented as random mixtures over latent topics and each topic defines a distribution over words. LDA assumes the following generative process for a corpus D consisting of M documents each of length N_i:

1. Choose $\theta_i \sim \text{Dir}(\alpha)$, where $i \in \{1,\ldots,M\}$ and $\text{Dir}(\alpha)$ is the Dirichlet distribution with parameter α
2. Choose $\varphi_k \sim \text{Dir}(\beta)$, where $k \in \{1,\ldots,K\}$
3. For each of the word positions i,j, where $j \in \{1,\ldots,N_i\}$, and $i \in \{1,\ldots,M\}$
   (a) Choose a topic $z_{i,j} \sim \text{Categorical}(\theta_i)$.
   (b) Choose a word $w_{i,j} \sim \text{Categorical}(\varphi_{z_{i,j}})$.

This is a smoothed LDA model to be precise. The subscript is often dropped, as in the plate diagrams 5. The aim is to compute the word probabilities of each topic, the topic of each word, and the particular topic mixture of each document. This can be done with Bayesian inference: the documents in the dataset represent the observations (evidence) and we want to compute the posterior distribution of the above quantities.

   This problem can modeled by the MCINTYRE program[7] below, where predicate

```
word(Doc,Position,Word)
```

indicates that document `Doc` in position `Position` (from 1 to the number of words of the document) has word `Word` and predicate

```
topic(Doc,Position,Topic)
```

---

[7] http://cplint.ml.unife.it/example/inference/lda.swinb.

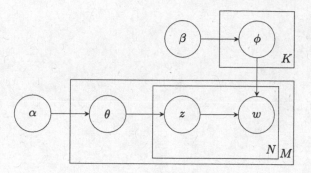

Fig. 5. Smoothed LDA.

indicates that document `Doc` associates topic `Topic` to the word in position `Position`. We also assume that the distributions for both θ_m and φ_k are symmetric Dirichlet distributions with scalar concentration parameter η set using a fact for the predicate `eta/1`, i.e., $\alpha = \beta = [\eta, \ldots, \eta]$. The program is then:

```
theta(_,Theta):dirichlet(Theta,Alpha):-
  alpha(Alpha).

topic(DocumentID,_,Topic):discrete(Topic,Dist):-
  theta(DocumentID,Theta),
  topic_list(Topics),
  maplist(pair,Topics,Theta,Dist).

word(DocumentID,WordID,Word):discrete(Word,Dist):-
  topic(DocumentID,WordID,Topic),
  beta(Topic,Beta),
  word_list(Words),
  maplist(pair,Words,Beta,Dist).

beta(_,Beta):dirichlet(Beta,Parameters):-
  n_words(N),
  eta(Eta),
  findall(Eta,between(1,N,_),Parameters).

alpha(Alpha):-
  eta(Eta),
  n_topics(N),
  findall(Eta,between(1,N,_),Alpha).

eta(2).

pair(V,P,V:P).
```

where `maplist/4` is a library of SWI-Prolog encoding the `maplist` primitive of functional programming. Suppose we have two topics, indicated with integers 1 and 2, and 10 words, indicated with integers $1, \ldots, 10$:

```
topic_list(L):-
  n_topics(N),
  numlist(1,N,L).

word_list(L):-
  n_words(N),
  numlist(1,N,L).

n_topics(2).

n_words(10).
```

We can, for example, use the model generatively and sample values for word in position 1 of document 1. The histogram of the frequency of word values when taking 100 samples is shown in Fig. 6.

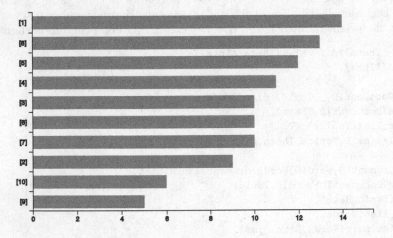

Fig. 6. Values for word in position 1 of document 1.eps

We can also sample values for couples (word, topic) in position 1 of document 1. The histogram of the frequency of the couples when taking 100 samples is shown in Fig. 7.

We can use the model to classify the words into topics. Here we use conditional inference with Metropolis-Hastings that is implemented in MCINTYRE. A priori both topics are about equally probable for word 1 of document, so if we take 100 samples of topic(1,1,T) we get the histogram in Fig. 8. If we observe that words 1 and 2 of document 1 are equal (word(1,1,1),word(1,2,1) as evidence) and take again 100 samples, one of the topics gets more probable, as the histogram of Fig. 9 shows. You can also see this if you look at the density of the probability of topic 1 before and after observing that words 1 and 2 of document 1 are equal: the observation makes the distribution less uniform, see Fig. 10.

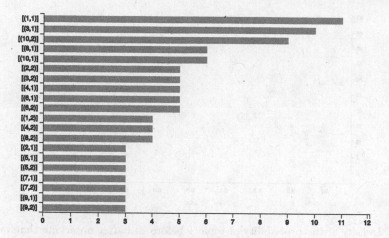

Fig. 7. Values for couples (word,topic) in position 1 of document 1.

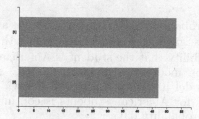

Fig. 8. Prior distribution of topics for word in position 1 of document 1.

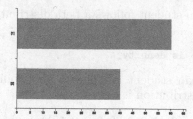

Fig. 9. Posterior distribution of topics for word in position 1 of document 1.

4.7   The Indian GPA Problem

In the Indian GPA problem proposed by Stuart Russel [28,29] the question is: if you observe that a student GPA is exactly 4.0, what is the probability that the student is from India, given that the American GPA score is from 0.0 to 4.0 and the Indian GPA score is from 0.0 to 10.0? Stuart Russel observed that most probabilistic programming system are not able to deal with this query because it requires combining continuous and discrete distributions. This problem can be modeled by building a mixture of a continuous and a discrete distribution for each nation to account for grade inflation (extreme values have a non-zero

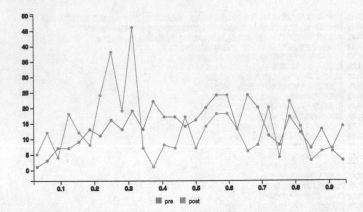

Fig. 10. Density of the probability of topic 1 before and after observing that words 1 and 2 of document 1 are equal.

probability). Then the probability of the student's GPA is a mixture of the nation mixtures. From statistics, given this model and the fact that the student's GPA is exactly 4.0, the probability that the student is American must be 1.0.

This problem can be modeled with Anglican, DC and MCINTYRE. In MCINTYRE we can model it with the program below[8]. The probability distribution of GPA scores for American students is continuous with probability 0.95 and discrete with probability 0.05:

```
is_density_A:0.95;is_discrete_A:0.05.
```

The GPA of an American student follows a beta distribution if the distribution is continuous:

```
agpa(A): beta(A,8,2) :- is_density_A.
```

The GPA of an American student is 4.0 with probability 0.85 and 0.0 with probability 0.15 if the distribution is discrete:

```
american_gpa(G) : finite(G,[4.0:0.85,0.0:0.15]) :- is_discrete_A.
```

or is obtained by rescaling the value of returned by `agpa/1` to the (0.0,4.0) interval:

```
american_gpa(A):- agpa(A0), A is A0*4.0.
```

The probability distribution of GPA scores for Indian students is continuous with probability 0.99 and discrete with probability 0.01.

```
is_density_I : 0.99; is_discrete_I:0.01.
```

The GPA of an Indian student follows a beta distribution if the distribution is continuous:

```
igpa(I): beta(I,5,5) :- is_density_I.
```

---

[8] http://cplint.ml.unife.it/example/inference/indian_gpa.pl.

The GPA of an Indian student is 10.0 with probability 0.9 and 0.0 with probability 0.1 if the distribution is discrete:

```
indian_gpa(I): finite(I,[0.0:0.1,10.0:0.9]):-  is_discrete_I.
```

or is obtained by rescaling the value returned by igpa/1 to the (0.0,10.0) interval:

```
indian_gpa(I) :- igpa(I0), I is I0*10.0.
```

The nation is America with probability 0.25 and India with probability 0.75.

```
nation(N) : finite(N,[a:0.25,i:0.75]).
```

The GPA of the student is computed depending on the nation:

```
student_gpa(G)  :- nation(a),american_gpa(G).
student_gpa(G)  :- nation(i),indian_gpa(G).
```

If we query the probability that the nation is America given that the student got 4.0 in his GPA we obtain 1.0, while the prior probability that the nation is America is 0.25.

5 Conclusions

PLP has now become mature enough to encode and solve a wide variety of problems. The recent inclusion of programs with infinite computation paths and continuous random variables closed the gap with other PP formalism, making PLP a valid alternative.

We have presented an overview of the semantics and of the main inference approaches, together with a set of examples that we believe show the maturity of the field.

Online tutorials on PLP are available at http://ds.ing.unife.it/~gcota/plptutorial/ [42] and https://dtai.cs.kuleuven.be/problog/tutorial.html.

References

1. Alberti, M., Bellodi, E., Cota, G., Riguzzi, F., Zese, R.: cplint on SWISH: probabilistic logical inference with a web browser. Intell. Artif. 11(1), 47–64 (2017)
2. Bellodi, E., Lamma, E., Riguzzi, F., Costa, V.S., Zese, R.: Lifted variable elimination for probabilistic logic programming. Theor. Pract. Log. Prog. 14(4–5), 681–695 (2014)
3. Bellodi, E., Riguzzi, F.: Experimentation of an expectation maximization algorithm for probabilistic logic programs. Intell. Artif. 8(1), 3–18 (2012)
4. Bellodi, E., Riguzzi, F.: Expectation maximization over binary decision diagrams for probabilistic logic programs. Intell. Data Anal. 17(2), 343–363 (2013)
5. Bellodi, E., Riguzzi, F.: Structure learning of probabilistic logic programs by searching the clause space. Theor. Pract. Log. Prog. 15(2), 169–212 (2015)
6. Blei, D.M., Ng, A.Y., Jordan, M.I.: Latent Dirichlet allocation. J. Mach. Learn. Res. 3, 993–1022 (2003)

7. Cozman, F.G., Mauá, D.D.: The structure and complexity of credal semantics. In: Hommersom, A., Abdallah, S.A. (eds.) PLP 2016, CEUR Workshop Proceedings, vol. 1661, pp. 3–14. CEUR-WS.org (2016)
8. Dantsin, E.: Probabilistic logic programs and their semantics. In: Voronkov, A. (ed.) RCLP -1990. LNCS, vol. 592, pp. 152–164. Springer, Heidelberg (1992). doi:10.1007/3-540-55460-2_11
9. Darwiche, A., Marquis, P.: A knowledge compilation map. J. Artif. Intell. Res. **17**, 229–264 (2002)
10. Raedt, L., Kersting, K.: Probabilistic inductive logic programming. In: Ben-David, S., Case, J., Maruoka, A. (eds.) ALT 2004. LNCS (LNAI), vol. 3244, pp. 19–36. Springer, Heidelberg (2004). doi:10.1007/978-3-540-30215-5_3
11. De Raedt, L., Kimmig, A.: Probabilistic (logic) programming concepts. Mach. Learn. **100**(1), 5–47 (2015)
12. De Raedt, L., Kimmig, A., Toivonen, H.: ProbLog: a probabilistic Prolog and its application in link discovery. In: Veloso, M.M. (ed.) IJCAI 2007, vol. 7, pp. 2462–2467. AAAI Press, Palo Alto (2007)
13. Di Mauro, N., Bellodi, E., Riguzzi, F.: Bandit-based Monte-Carlo structure learning of probabilistic logic programs. Mach. Learn. **100**(1), 127–156 (2015)
14. Fierens, D., Van den Broeck, G., Renkens, J., Shterionov, D.S., Gutmann, B., Thon, I., Janssens, G., De Raedt, L.: Inference and learning in probabilistic logic programs using weighted boolean formulas. Theor. Pract. Log. Prog. **15**(3), 358–401 (2015)
15. Gorlin, A., Ramakrishnan, C.R., Smolka, S.A.: Model checking with probabilistic tabled logic programming. Theor. Pract. Log. Prog. **12**(4–5), 681–700 (2012)
16. Gutmann, B., Thon, I., Kimmig, A., Bruynooghe, M., De Raedt, L.: The magic of logical inference in probabilistic programming. Theor. Pract. Log. Prog. **11**(4–5), 663–680 (2011)
17. Holzinger, A.: Introduction to machine learning and knowledge extraction (MAKE). Mach. Learn. Knowl. Extr. **1**(1), 1–20 (2017)
18. Holzinger, A., Schantl, J., Schroettner, M., Seifert, C., Verspoor, K.: Biomedical text mining: state-of-the-art, open problems and future challenges. In: Holzinger, A., Jurisica, I. (eds.) Interactive Knowledge Discovery and Data Mining in Biomedical Informatics. LNCS, vol. 8401, pp. 271–300. Springer, Heidelberg (2014). doi:10.1007/978-3-662-43968-5_16
19. Hurd, J.: A formal approach to probabilistic termination. In: Carreño, V.A., Muñoz, C.A., Tahar, S. (eds.) TPHOLs 2002. LNCS, vol. 2410, pp. 230–245. Springer, Heidelberg (2002). doi:10.1007/3-540-45685-6_16
20. Islam, M.A., Ramakrishnan, C., Ramakrishnan, I.: Inference in probabilistic logic programs with continuous random variables. Theor. Pract. Log. Prog. **12**, 505–523 (2012)
21. Kaminski, B.L., Katoen, J.-P., Matheja, C., Olmedo, F.: Weakest precondition reasoning for expected run–times of probabilistic programs. In: Thiemann, P. (ed.) ESOP 2016. LNCS, vol. 9632, pp. 364–389. Springer, Heidelberg (2016). doi:10.1007/978-3-662-49498-1_15
22. Kilgour, D.M., Brams, S.J.: The truel. Math. Mag. **70**(5), 315–326 (1997)
23. Kimmig, A., Demoen, B., De Raedt, L., Costa, V.S., Rocha, R.: On the implementation of the probabilistic logic programming language ProbLog. Theor. Pract. Log. Prog. **11**(2–3), 235–262 (2011)

24. Meert, W., Struyf, J., Blockeel, H.: CP-logic theory inference with contextual variable elimination and comparison to BDD based inference methods. In: Raedt, L. (ed.) ILP 2009. LNCS (LNAI), vol. 5989, pp. 96–109. Springer, Heidelberg (2010). doi:10.1007/978-3-642-13840-9_10

25. Muggleton, S.: Learning stochastic logic programs. Electron. Trans. Artif. Intell. **4**(B), 141–153 (2000)

26. Nampally, A., Ramakrishnan, C.: Adaptive MCMC-based inference in probabilistic logic programs. arXiv preprint arXiv:1403.6036 (2014)

27. Ng, R.T., Subrahmanian, V.S.: Probabilistic logic programming. Inf. Comput. **101**(2), 150–201 (1992)

28. Nitti, D., De Laet, T., De Raedt, L.: Probabilistic logic programming for hybrid relational domains. Mach. Learn. **103**(3), 407–449 (2016)

29. Perov, Y., Paige, B., Wood, F.: The Indian GPA problem (2017). http://www.robots.ox.ac.uk/~fwood/anglican/examples/viewer/?worksheet=indian-gpa. Accessed 15 Apr 2017

30. Pfeffer, A.: Practical Probabilistic Programming. Manning Publications, Cherry Hill (2016)

31. Poole, D.: The independent choice logic for modelling multiple agents under uncertainty. Artif. Intell. **94**, 7–56 (1997)

32. Poole, D.: Abducing through negation as failure: stable models within the independent choice logic. J. Logic Program. **44**(1–3), 5–35 (2000)

33. Poole, D.: Probabilistic horn abduction and Bayesian networks. Artif. Intell. **64**(1), 81–129 (1993)

34. Richardson, M., Domingos, P.: Markov logic networks. Mach. Learn. **62**(1–2), 107–136 (2006)

35. Riguzzi, F.: ALLPAD: approximate learning of logic programs with annotated disjunctions. Mach. Learn. **70**(2–3), 207–223 (2008)

36. Riguzzi, F.: MCINTYRE: a Monte Carlo system for probabilistic logic programming. Fund. Inform. **124**(4), 521–541 (2013)

37. Riguzzi, F.: Speeding up inference for probabilistic logic programs. Comput. J. **57**(3), 347–363 (2014)

38. Riguzzi, F.: The distribution semantics for normal programs with function symbols. Int. J. Approx. Reason. **77**, 1–19 (2016)

39. Riguzzi, F., Bellodi, E., Lamma, E., Zese, R., Cota, G.: Probabilistic logic programming on the web. Softw. Pract. Exper. **46**(10), 1381–1396 (2016)

40. Riguzzi, F., Bellodi, E., Zese, R., Cota, G., Lamma, E.: Scaling structure learning of probabilistic logic programs by mapreduce. In: Fox, M., Kaminka, G. (eds.) ECAI 2016, vol. 285, pp. 1602–1603. IOS Press, Prague (2016)

41. Riguzzi, F., Bellodi, E., Zese, R., Cota, G., Lamma, E.: A survey of lifted inference approaches for probabilistic logic programming under the distribution semantics. Int. J. Approx. Reason. **80**, 313–333 (2017)

42. Riguzzi, F., Cota, G.: Probabilistic logic programming tutorial. Assoc. Log. Program. Newslett. **29**(1), 1–1 (2016)

43. Riguzzi, F., Di Mauro, N.: Applying the information bottleneck to statistical relational learning. Mach. Learn. **86**(1), 89–114 (2012)

44. Riguzzi, F., Lamma, E., Alberti, M., Bellodi, E., Zese, R., Cota, G.: Probabilistic logic programming for natural language processing. In: Chesani, F., Mello, P., Milano, M. (eds.) Workshop on Deep Understanding and Reasoning, URANIA 2016, CEUR Workshop Proceedings, vol. 1802, pp. 30–37 (2017)

45. Riguzzi, F., Swift, T.: The PITA system: tabling and answer subsumption for reasoning under uncertainty. Theor. Pract. Log. Prog. **11**(4–5), 433–449 (2011)

46. Riguzzi, F., Swift, T.: Welldefinedness and efficient inference for probabilistic logic programming under the distribution semantics. Theor. Pract. Log. Prog. **13**(Special Issue 02–25th Annual GULP Conference), 279–302 (2013)

47. Riguzzi, F., Zese, R., Cota, G.: Probabilistic inductive logic programming on the web. In: Ciancarini, P., Poggi, F., Horridge, M., Zhao, J., Groza, T., Suarez-Figueroa, M.C., d'Aquin, M., Presutti, V. (eds.) EKAW 2016. LNCS (LNAI), vol. 10180, pp. 172–175. Springer, Cham (2017). doi:10.1007/978-3-319-58694-6_25

48. Sato, T., Meyer, P.: Infinite probability computation by cyclic explanation graphs. Theor. Pract. Log. Prog. **14**, 909–937 (2014)

49. Sato, T.: A statistical learning method for logic programs with distribution semantics. In: Sterling, L. (ed.) ICLP 1995, pp. 715–729. MIT Press, Cambridg (1995)

50. Sato, T., Kameya, Y.: PRISM: a language for symbolic-statistical modeling. In: IJCAI 1997, vol. 97, pp. 1330–1339 (1997)

51. Sato, T., Kubota, K.: Viterbi training in prism. Theor. Pract. Log. Prog. **15**(02), 147–168 (2015)

52. Sato, T., Meyer, P.: Tabling for infinite probability computation. In: Dovier, A., Costa, V.S. (eds.) ICLP TC 2012. LIPIcs, vol. 17, pp. 348–358 (2012)

53. Valiant, L.G.: The complexity of enumeration and reliability problems. SIAM J. Comput. **8**(3), 410–421 (1979)

54. Van Gelder, A., Ross, K.A., Schlipf, J.S.: The well-founded semantics for general logic programs. J. ACM **38**(3), 620–650 (1991)

55. Vennekens, J., Verbaeten, S.: Logic programs with annotated disjunctions. Technical report CW386, KU Leuven (2003)

56. Vennekens, J., Verbaeten, S., Bruynooghe, M.: Logic programs with annotated disjunctions. In: Demoen, B., Lifschitz, V. (eds.) ICLP 2004. LNCS, vol. 3132, pp. 431–445. Springer, Heidelberg (2004). doi:10.1007/978-3-540-27775-0_30

57. Von Neumann, J.: Various techniques used in connection with random digits. Nat. Bureau Stand. Appl. Math. Ser. **12**, 36–38 (1951)

58. Wielemaker, J., Schrijvers, T., Triska, M., Lager, T.: SWI-Prolog. Theor. Pract. Log. Prog. **12**(1–2), 67–96 (2012)

Persistent Topology for Natural Data Analysis — A Survey

Massimo Ferri[✉]

Dip. di Matematica and ARCES, Univ. di Bologna, Bologna, Italy
massimo.ferri@unibo.it

Abstract. Natural data offer a hard challenge to data analysis. One set of tools is being developed by several teams to face this difficult task: Persistent topology. After a brief introduction to this theory, some applications to the analysis and classification of cells, liver and skin lesions, music pieces, gait, oil and gas reservoirs, cyclones, galaxies, bones, brain connections, languages, handwritten and gestured letters are shown.

Keywords: Homology · Betti numbers · Size functions · Filtering function · Classification · Retrieval

1 Introduction and Motivation

What is the particular challenge offered by natural data, which could suggest the need of topology, and in particular of persistence? Simply said, it's quality instead of quantity. This is especially evident with images.

If one has to analyze, classify, retrieve images of mechanical pieces, vehicles, rigid objects, then geometry fulfills all needs. On the images themselves, matrix theory provides the transformations for superimposing a picture to a template. More often, pictures are represented by feature vectors, whose components are geometric measures (*shape descriptors*). Then recognition, defect detection, retrieval etc. can be performed on the feature vectors.

The scene changes if the depicted objects are of natural origin: the rigidity of geometry becomes an obstacle. Recognizing the resemblance between a sitting and a standing man is difficult. The challenge is even harder when it comes to biomedical data and when the context is essential for the understanding of data [34,51].

It's here that topology comes into play: the standing and sitting men are *homeomorphic*, i.e. there is a topological transformation which superimposes one to the other, whereas no matrix will ever be able to do that. It is generally difficult to discover whether two objects are homeomorphic; then algebraic topology turns helpful: It associates invariants — e.g. Betti numbers — to topological spaces, such that objects which are homeomorphic have identical invariants (the converse does not hold, unfortunately).

(Algebraic) topology seems then to be the right environment for formalizing qualitative aspects in a computable way, as is nicely expressed in [35, Sect. 5.1].

A. Holzinger et al. (Eds.): Integrative Machine Learning, LNAI 10344, pp. 117–133, 2017.
https://doi.org/10.1007/978-3-319-69775-8_6

There is a problem: if geometry is too rigid, topology is too free. This is the reason why persistent topology can offer new topological descriptors (e.g. Persistent Betti Numbers, Persistence Diagrams) which preserve some selected geometric features through *filtering functions*. Classical references on persistence are [8,12,22,50].

Persistent topology has been experimented in the image context, particularly in the biomedical domain, but also in fields where data are not pictures, e.g. in geology, music and linguistics, as will be shown in this survey.

2 Glossary and Basic Notions

It is out of the scope of this survey to give a working introduction to homology and persistence; we limit ourselves to an intuitive description of the concepts, and recommend to profit of the technical references, without which a real understanding of the results is impossible. An essential (and avoidable) technical description of a particular homology is reported in Sect. 2.1.

Homology. There is a well-structured way (technically a set of functors) to associate *homology vector spaces* (more generally modules) $H_k(X)$ to a simplicial complex or to a topological space X, and linear transformations to maps [33, Chap. 2] and [23, Chap. 4].

Betti numbers. The *k-th Betti number* $\beta_k(X)$ is the dimension of the k-th homology vector space $H_k(X)$, i.e. the number of independent generators (homology classes of k-*cycles*) of this space. Intuitively, $\beta_0(X)$ counts the number of path-connected components (i.e. the separate pieces) of which X is composed; $\beta_1(X)$ counts the holes of the type of a circle (like the one of a doughnut); $\beta_2(X)$ counts the 2-dimensional voids (like the ones of gruyere or of an air chamber).

Homeomorphism. Given topological spaces X and Y, a *homeomorphism* from X to Y is a continuous map with continuous inverse. If one exists, the two spaces are said to be *homeomorphic*. This is the typical equivalence relation between topological spaces. Homology vector spaces and Betti numbers are invariant under homeomorphisms.

Remark 1. As hinted in the Introduction, geometry is too rigid, but topology is too free. In particular, homeomorphic spaces can be very different from an intuitive viewpoint: the joke by which "for a topologist a mug and a doughnut are the same" is actually true; the two objects are homeomorphic! **Persistent topology** then tries to overcome this difficulty by studying not just topological spaces but pairs, once called *size pairs*, (X, f) where f is generally a continuous function, called *measuring* or *filtering function*, from X to $\mathbb{R}$ (to $\mathbb{R}^n$ in *multidimensional persistence*) which conveys the idea of shape, the viewpoint of the observer. Shape similarity is actually very much dependent on the context. The Betti numbers of the sublevel sets then make it possible to distinguish the two objects although they are homeomorphic: see Fig. 1.

Fig. 1. Sublevel sets of mug and doughnut.

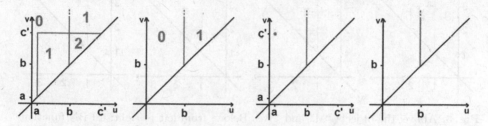

Fig. 2. From left to right: 1-PBN functions of mug and of doughnut, 1-PDs of mug and of doughnut.

Sublevel Sets. Given a pair (X, f), with $f : X \to \mathbb{R}$ continuous, given $u \in \mathbb{R}$, the *sublevel set under u* is the set $X_u = \{x \in X \mid f(x) \leq u\}$.

Persistent Betti Numbers. For all $u, v \in \mathbb{R}$, $u < v$, the inclusion map $\iota^{u,v} : X_u \to X_v$ is continuous and induces, at each degree k, a linear transformation $\iota_*^{u,v} : H_k(X_u) \to H_k(X_v)$. The *$k$-Persistent Betti Number (k-PBN) function* assigns to the pair (u, v) the number $\dim \operatorname{Im} \iota_*^{u,v}$, i.e. the number of classes of k-cycles of $H_k(X_u)$ which "survive" in $H_k(X_v)$. See Fig. 2 (left) for the 1-PBN functions of mug and doughnut. Note that a pitcher, and more generally any open container with a handle, will have very similar PBNs to the ones of the mug; this is precisely what we want for a functional search and not for a strictly geometrical one.

Persistence Diagrams. The k-PBN functions are wholly determined by the position of some discontinuity points and lines, called *cornerpoints* and *cornerlines* (or *cornerpoints at infinity*) The coordinates (u, v) of a cornerpoint represent the levels of "birth" and "death" respectively of a generator; the abscissa of a cornerline is the level of birth of a generator which never dies. The *persistence* of a cornerpoint is the difference $v - u$ of its coordinates. Cornerpoints and cornerlines form the *k-Persistence Diagram (k-PD)*. Figure 2 (right) depicts the 1-PDs of mug and of doughnut. For the sake of simplicity, we are here neglecting the fact that cornerpoints and cornerlines may have multiplicities.

Remark 2. Sometimes it is important to distinguish even objects for which there exists a rigid movement superimposing one to the other — so also geometrically equivalent — as in the case of some letters: context may be essential! See Fig. 3, where ordinate plays the role of filtering function.

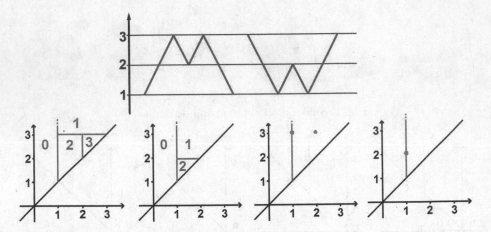

Fig. 3. Above: the objects "M" and "W". Below, from left to right: 0-PBN functions of M and of W, 0-PDs of M and of W.

Matching distance. Given the k-PDs $\mathcal{D}_{X,f}, \mathcal{D}_{Y,g}$ of two pairs $(X, f), (Y, g)$, match the cornerpoints of $\mathcal{D}_{X,f}$ either with cornerpoints of $\mathcal{D}_{Y,g}$ or with their own projections on the diagonal $u = v$; the *weight* of this matching is the sup of the L_∞-distances of matching points. The *matching distance* (or *bottleneck distance*) of $\mathcal{D}_{X,f}$ and $\mathcal{D}_{Y,g}$ is the inf of such weights among all possible such matchings.

Natural pseudodistance. Given two pairs $(X, f), (Y, g)$, with X, Y homeomorphic, the *weight* of a given homeomorphism $\varphi : X \to Y$ is $\sup_{x \in X} |g(\varphi(x)) - f(x)|$. The *natural pseudodistance* of (X, f) and (Y, g) is the inf of these weights among all possible homeomorphisms. If we are given the k-PDs of the two pairs, their matching distance is a lower bound for the natural pseudodistance of the two pairs, and it is the best possible obtainable from the two k-PDs. Much is known on this dissimilarity measure [19–21].

2.1   A Brief Technical Description of Homology

There are several homologies. The classical and most descriptive one, at least for compact spaces, is singular homology with coefficients in $\mathbb{Z}$; we refer to [33, Chap. 2] for a thorough exposition of it. Anyway, the homology used in most applications is the simplicial one, of which (with coefficients in $\mathbb{Z}_2$) we now give a very short introduction following [23, Chap. 4].

Simplices. A *p-simplex* σ is the convex hull, in a Euclidean space, of a set of $p + 1$ points, called *vertices* of the simplex, not contained in a Euclidean $(p-1)$-dimensional subspace; the simplex is said to be *generated* by its vertices. A *face* of a simplex σ is the simplex generated by a nonempty set of vertices of σ.

Simplicial complexes. A finite collection K of simplices of a given Euclidean space is a *simplicial complex* if (1) for any $\sigma \in K$, all faces of σ belong to K,

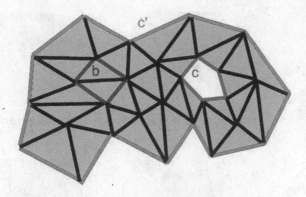

Fig. 4. Cycles.

(2) the intersection of two simplices of K is either empty or a common face. The *space* of the complex K is the topological subspace of Euclidean space $|K|$ formed by the union of all simplices of K.

Simplicial homology with $\mathbb{Z}_2$ coefficients. Given a (finite) simplicial complex K, call *p-chain* any formal linear combination of p-simplices with coefficients in $\mathbb{Z}_2$ (i.e. either 1 or 0, with $1 + 1 = 0$). p-chains form a $\mathbb{Z}_2$-vector space C_p. Note that each p-chain actually identifies a set of p-simplices of K and that the sum of two p-chains is just the symmetric difference (Xor) of the corresponding sets. We now introduce a linear transformation $\partial_p : C_p \to C_{p-1}$ (called *boundary operator*) for any $p \in \mathbb{Z}$. We just need to define it on generators, i.e. on p-simplices, and then extend by linearity. Writing $\sigma = [u_0, u_1, \ldots, u_p]$, we denote by $[u_0, \ldots, \hat{u}_j, \ldots, u_p]$ its face generated by all of its vertices except u_j $(j = 0, \ldots, p)$. Then we define

$$\partial_p(\sigma) = \sum_{j=0}^{n} [u_0, \ldots, \hat{u}_j, \ldots, u_p]$$

It is possible to prove that $\partial_p \partial_{p+1} = 0$, so that $B_p = \text{Im} \partial_{p+1}$ is contained in $Z_p = \text{Ker} \partial_p$. Elements of B_p are called *p-boundaries*; elements of Z_p are called *p-cycles*. The *p-homology vector space* is defined as the quotient $H_p(K) = Z_p/B_p$. Homology classes are represented by cycles which are not boundaries. Two cycles are *homologous* is their difference is a boundary. In Fig. 4, representing the simplicial complex K formed by the shaded triangles and their faces, the blue chain b is a 1-cycle which is also a boundary; the red chain c and the green one c' are 1-cycles which are not boundaries; c and c' are homologous.

3 State-of-the-Art

The application of persistence to shape analysis and classification has a long story, since it started in the 90's when it still had the name of Size Theory [50]. In the last few years it has taken various, very interesting forms. The constant

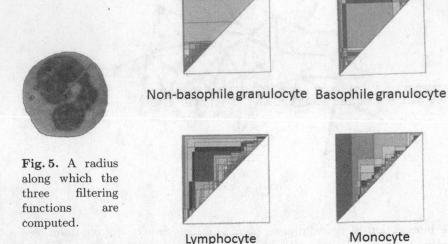

Non-basophile granulocyte   Basophile granulocyte

Fig. 5. A radius along which the three filtering functions are computed.

Lymphocyte   Monocyte

Fig. 6. Persistent Betti Number functions relative to the sum of grey tones (different colors represent different values).

aspect is always the presence of qualitative features which are difficult to capture and formalize within other frames of mind.

3.1 Leukocytes

Leukocytes, or white blood cells, belong to five different classes: lymphocyte; monocyte; neutrophile, eosinophile, basophile granulocytes. Eosinophile and neutrophile granulocytes are generally difficult to be distinguished, so they were considered in a single classification class in an early research by the Bologna team [26].

As a space, the boundary of the starlike hull of the cell is assumed. The images are converted to grey tones.

Three filtering functions are put to work, all computed along radii from the center of mass of the cell (Fig. 5):

– Sum of grey tones
– Maximum variation
– Sum of variations pixel to pixel.

Classification (with very good hit ratios for that time) is performed by measuring distance from the average PBN function of each class.

3.2 Handwritten Letters and Monograms

Again in Bologna we faced recognition of handwritten letters with time information; our goal was to recognize both the alphabet letter and the writer [25].

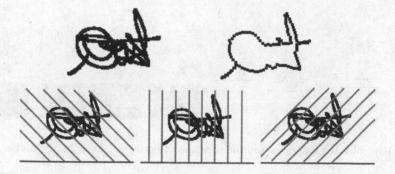

Fig. 7. A monogram with its outline (above) and the directions along which the filtering functions are computed (below).

The space on which the filtering functions are defined is the time interval of the writing. The filtering functions are computed in the 3D "plane-time":

- Distance of points from the letter axis
- Speed
- Curvature
- Torsion
- Distance from center of mass (in plane projection).

Classification comes from fuzzy characteristic functions, obtained from normalized inverse of distance. Cooperation of the characteristic functions coming from the single filtering functions is given by their rough arithmetic average.

A later experiment, which was even repeated live at a conference, concerned the recognition of monograms for personal identification, without time information [24].

Two topological spaces are used. The first is the outline of the monogram and the filtering function is the distance from the center of mass (see upper Fig. 7).

The second space is a horizontal segment placed at the base of the monogram image. Filtering functions:

- Number of black pixels along segments (3 directions) (see lower Fig. 7)
- Number of pixel-pixel black-white jumps (3 directions).

Classification is performed by a weighted average of fuzzy characteristic functions.

3.3 Sign Alphabet

Automatic recognition of the symbols expressed by the hands in the sign language is a task which was of interest for different teams. The first one was the group led by Alessandro Verri in Genova [49]. The signs were performed with a

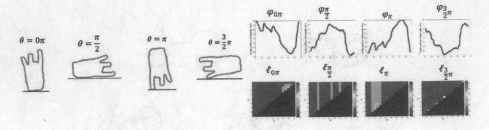

Fig. 8. Four filtering functions and the corresponding 0-Persistent Betti Number functions.

Fig. 9. Four filtering functions on silhouette stacks for gait identification.

white glove on a black background; translation into common letters was done in real time in a live demo at a conference.

The domain space is a horizontal segment; the filtering functions assign to each point of the segment the maximum distance of a contour point within a strip of fixed width, with 24 different strip orientations.

The choice of S. Wang in Sherbrooke, instead, is to use a part of the contour, determined by principal component analysis, as a domain and distance from center of mass as filtering function [32].

The team of D.Kelly in Maynooth uses the whole contour as domain, and distances from four lines as filtering functions [36] (see Fig. 8).

3.4 Human Gait

Personal identification and surveillance are the aim of a research by the Cuban team of L. Lamar-León, together with the Sevilla group of computational topology [37].

Considering a stack of silhouettes as a 3D object, and using four different filtering functions, makes 0- and 1-degree persistent homology a tool for identifying people through their gait (Fig. 9).

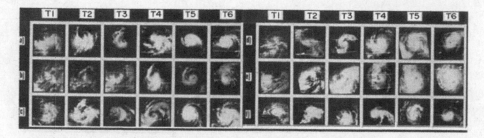

Fig. 10. Time evolution of cyclones.

3.5 Tropical Cyclones

S. Banerjee in Kolkata makes use of persistence on sequences of satellite images of cloud systems (Fig. 10), in order to evaluate risk and intensity of forming hurricanes [2].

Time interval is the domain of two filtering functions which are common characteristic measures of cyclones:

– Central Feature portion
– Outer Banding Feature

3.6 Galaxies

Again S. Banerjee [3] applies similar methods to another type of spirals: galaxies.

Various filtering functions are used. One is defined as a function of distance from galaxy center, and is the ratio between major and minor axis of the corresponding isophote. Another one is a "pitch" parameter defined by Ringermacher and Mead [45]. A third filtering function is a compound based on color.

The classification results agree with the literature.

3.7 Bones

In [48] a powerful construction (the Persistent Homology Transform) is introduced. It consists in gathering the "height" filtering functions according to all possible directions. The paper shows that the transform is injective for objects homeomorphic to spheres. By using the transform it is possible to define an effective distance between surfaces. An application is shown by classifying heel bones of different species; the comparison with the ground truth produced by using placement of landmarks on the surfaces is very good.

3.8 Melanocytic Lesions

A very important part of natural shape analysis is the detection of malignant cells and lesions, since there generally are no templates for them. As far as we

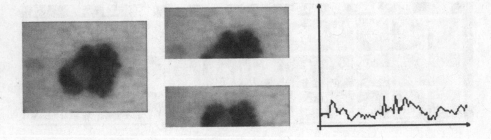

Fig. 11. One of the 45 splittings of a melanocytic lesion, and the whole A-curve corresponding to the filtering function luminance.

know, the first attempt through persistence (called *size theory* at that time) is the ADAM EU Project, by the Bologna team together with CINECA and with I. Stanganelli, a dermatologist of the Romagna Oncology Institute [17,27,47]. The analysis is mainly based on asymmetry of boundary, masses and color distribution: the lesion is split into two halves by 45 equally spaced lines, and the difference between the two halves is measured by the matching distance of the corresponding Persistence Diagrams.

The three functions (A-curves) relating these distances to the splitting line angles give parameters which are then fed into a Support Vector Machine classifier.

The same team is presently involved with a biomedical firm in the realization of a machine for smart retrieval of dermatological images [28].

3.9   Tumor Mouth Cells

A morphological classification of normal and tumor cells of the epithelial tissue of the mouth is proposed in [40,41]: the filtering function is distance from the center of mass; the discrimination is statistically based on the distribution of cornerpoints (see Fig. 12).

3.10   Hepatic Lesions

The advantages of a multidimensional range for the filtering functions are shown in [1], where several classification experiments are performed on the images of hepatic cells (see Fig. 13). The domain space is the part of image occupied by the lesion; the two components of the filtering function are the greyscale of each pixel and the distance from the lesion boundary.

3.11   Genetic Pathways

So far we have seen applications of persistence to images of natural origin. But the modularity of the method opens the possibility to deal with data of very different nature. A first example is given by [43], where persistence is used on the

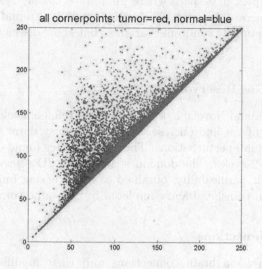

Fig. 12. Distribution of cornerpoints in the diagrams of normal and tumor mouth cells.

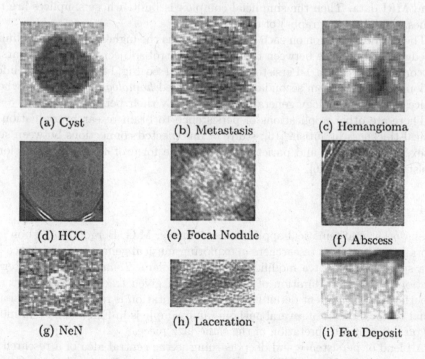

(a) Cyst

(b) Metastasis

(c) Hemangioma

(d) HCC

(e) Focal Nodule

(f) Abscess

(g) NeN

(h) Laceration

(i) Fat Deposit

Fig. 13. Various types of hepatic lesions.

Vietoris-Rips complex in a space where points are *complex phenotypes* related together by the *Jaccard distance*. This made it possible to find systematic associations among metabolic syndrome variates that show distinctive genetic association profiles.

3.12 Oil and Gas Reservoirs

Researchers in Ufa and Novosibirsk need to get a reliable geological and hydro-dynamical model of gas and oil reservoirs out of noisy data; the model has to be robust under small perturbations. The authors have found an answer in persistent 0-, 1- and 2-cycles. The domain space is the 3D reservoir bed, and the filtering function is permeability, obtained as a decreasing function of radioactivity [4] (Russian; translated and completed in this same volume).

3.13 Brain Connections

A complex research on brain connections and their modification under the assumption of a psychoactive substance (psilocybine) is performed in [42] and extended in [39]. The construction starts with a complete graph whose vertices are cortical or subcortical regions; these, and their functional connectivity (expressed as weights on the edges) come from an elaborate processing of functional MRI data. Then the simplicial complex is built, whose simplices are the cliques (complete subgraphs) of the graph.

The filtering function on each simplex is minus the highest weight of its building edges. A difference between treated and control subjects already appears in the comparison of the 1-Persistence Diagrams (see Fig. 14). Then more information is obtained from secondary graphs (called *homological scaffolds*), whose vertices are the homology generators weighted by their persistence.

There are other applications of persistence to brain research: evaluation of cortical thickness in autism [16]; study of unexpected connections between subcortex, frontal cortex and parietal cortex in the form of 1- and 2-dimensional persistent cycles [31, 46].

3.14 Music

Among other mathematical applications to music, M.G. Bergomi in Lisbon collaborates with various researchers in exploring musical genres by persistence [6]. As a space they adopt a modified version of Euler's *Tonnetz* [9]. The filtering function is the total duration of each note in a given track. Classification can be performed at different detail levels: experimentation is reported on tonal and atonal classical music of several authors (an example is in Fig. 15), on pop music and on different interpretation of the same jazz piece.

A blend of persistence and deep learning is the central idea of a research by the team of I.-H. Yang in Taiwan [38]. They input audio signals to a Convolutional Neural Network (CNN); after a first convolution layer, a middle layer

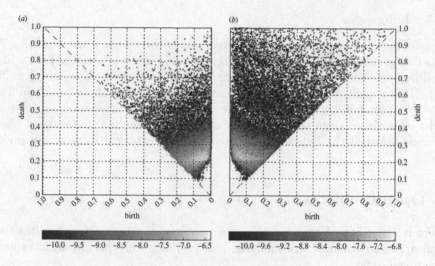

Fig. 14. Probability densities for H_1 generators: placebo (left) and psilocybin (right) treated.

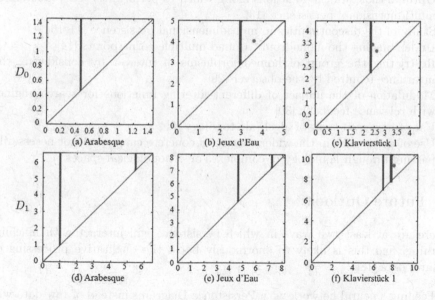

Fig. 15. 0- and 1-persistence diagrams for three classical pieces.

processes the output of the first in two different complementary ways: one is a classical CNN; the other computes the persistence landscape (an information piece derivable from the persistence diagram [10]) of the same output. Whereas the persistence layer by itself does not perform any better than the normal CNN, their combination gives very good results in terms of music tagging.

3.15 Languages

An interdisciplinary team at Caltech investigates the metric spaces built by 79 Indo-European and 49 Niger-Congo languages [44]. These appear as points in a Euclidean space of syntactic parameters; on them a Vietoris-Rips complex [23, Sect. III.2] is built and Euclidean distance is assumed as filtering function. The Indo-European family reveals one 1-dimensional and two 0-dimensonal persistent cycles, the Niger-Congo respectively none and one. The interpretation of these differences and of the link with phylogenetic and historical facts is still under way.

4   Open Problems

There is a number of open problems in persistence, whose solution will affect applications to natural data analysis, and to which only partial answers have been given so far:

- Optimal choice of the foliations along which to perform the 1D reduction of multidimensional persistence [13]
- Study of the discontinuities in multidimensional persistence [11,15]
- Understanding the monodromy around multiple cornerpoints [14]
- Restricting the group of homeomorphisms of interest by considering the invariance required by the observer [29]
- Modulation of the impact of different filtering functions for search engines with relevance feedback [30]
- Use of advanced tools of algebraic topology [5]
- Use of persistence in the wider context of concrete categories, not necessarily passing through homology of complexes or of topological spaces [7].

5   Future Outlook

There are at least two ways in which persistence will interact with machine learning, and this is likely to enormously boost the qualitative processing of natural data [18]:

- Feeding a neural network with Persistence Diagrams instead of raw data will convey the needs and viewpoints of the user
- Deep learning might yield a quantum leap in persistence, by automatically finding the best filtering functions for a given problem.

Acknowledgments. Article written within the activity of INdAM-GNSAGA.

References

1. Adcock, A., Rubin, D., Carlsson, G.: Classification of hepatic lesions using the matching metric. Comput. Vis. Image Underst. **121**, 36–42 (2014)
2. Banerjee, S.: Size functions in the study of the evolution of cyclones. Int. J. Meteorol. **36**(358), 39 (2011)
3. Banerjee, S.: Size functions in galaxy morphology classification. Int. J. Comput. Appl. **100**(3), 1–4 (2014)
4. Bazaikin, Y.V., Baikov, V.A., Taimanov, I.A., Yakovlev, A.A.: Chislennyi analiz topologicheskih harakteristik trehmernyh geologicheskih modelei neftegazovyh mestorozhdenii. Matematicheskoe Modclirovanie **25**(10), 19–31 (2013)
5. Belchí, F., Murillo, A.: A_∞-persistence. Appl. Algebra Eng. Commun. Comput. **26**(1–2), 121–139 (2015)
6. Bergomi, M.G., Baratè, A., Di Fabio, B.: Towards a topological fingerprint of music. In: Bac, A., Mari, J.I. (eds.) CTIC 2016. LNCS, vol. 9667, pp. 88–100. Springer, Cham (2016). doi:10.1007/978-3-319-39441-1_9
7. Bergomi, M.G., Ferri, M., Zuffi, L.: Graph persistence. arXiv preprint arXiv:1707.09670 (2017)
8. Biasotti, S., Cerri, A., Frosini, P., Giorgi, D., Landi, C.: Multidimensional size functions for shape comparison. J. Math. Imag. Vis. **32**(2), 161–179 (2008)
9. Bigo, L., Andreatta, M., Giavitto, J.-L., Michel, O., Spicher, A.: Computation and visualization of musical structures in chord-based simplicial complexes. In: Yust, J., Wild, J., Burgoyne, J.A. (eds.) MCM 2013. LNCS (LNAI), vol. 7937, pp. 38–51. Springer, Heidelberg (2013). doi:10.1007/978-3-642-39357-0_3
10. Bubenik, P., Dłotko, P.: A persistence landscapes toolbox for topological statistics. J. Symb. Comput. **78**, 91–114 (2017)
11. Carlsson, G., Zomorodian, A.: The theory of multidimensional persistence. Discr. Comput. Geom. **42**(1), 71–93 (2009)
12. Carlsson, G., Zomorodian, A., Collins, A., Guibas, L.J.: Persistence barcodes for shapes. IJSM **11**(2), 149–187 (2005)
13. Cerri, A., Di Fabio, B., Ferri, M., Frosini, P., Landi, C.: Betti numbers in multidimensional persistent homology are stable functions. Math. Methods Appl. Sci. **36**(12), 1543–1557 (2013)
14. Cerri, A., Ethier, M., Frosini, P.: A study of monodromy in the computation of multidimensional persistence. In: Gonzalez-Diaz, R., Jimenez, M.-J., Medrano, B. (eds.) DGCI 2013. LNCS, vol. 7749, pp. 192–202. Springer, Heidelberg (2013). doi:10.1007/978-3-642-37067-0_17
15. Cerri, A., Frosini, P.: Necessary conditions for discontinuities of multidimensional persistent betti numbers. Math. Methods Appl. Sci. **38**(4), 617–629 (2015)
16. Chung, M.K., Bubenik, P., Kim, P.T.: Persistence diagrams of cortical surface data. In: Prince, J.L., Pham, D.L., Myers, K.J. (eds.) IPMI 2009. LNCS, vol. 5636, pp. 386–397. Springer, Heidelberg (2009). doi:10.1007/978-3-642-02498-6_32
17. d'Amico, M., Ferri, M., Stanganelli, I.: Qualitative asymmetry measure for melanoma detection. In: IEEE International Symposium on Biomedical Imaging: Nano to Macro, pp. 1155–1158. IEEE (2004)
18. Dehmer, M., Emmert-Streib, F., Pickl, S., Holzinger, A.: Big Data of Complex Networks. CRC Press, Boca Raton (2016)
19. Donatini, P., Frosini, P.: Lower bounds for natural pseudodistances via size functions. Arch. Inequal. Appl. **1**(2), 1–12 (2004)

20. Donatini, P., Frosini, P.: Natural pseudodistances between closed manifolds. Forum Mathematicum **16**(5), 695–715 (2004)
21. Donatini, P., Frosini, P.: Natural pseudodistances between closed surfaces. J. Eur. Math. Soc. **9**(2), 231–253 (2007)
22. Edelsbrunner, H., Harer, J.: Persistent homology–a survey. In: Surveys on Discrete and Computational Geometry, vol. 453, pp. 257–282, Providence, RI (2008). Contemp. Math. Amer. Math. Soc
23. Edelsbrunner, H., Harer, J.: Computational Topology: An Introduction. American Mathematical Society, Providence (2009)
24. Ferri, M., Frosini, P., Lovato, A., Zambelli, C.: Point selection: a new comparison scheme for size functions (with an application to monogram recognition). In: Chin, R., Pong, T.-C. (eds.) ACCV 1998. LNCS, vol. 1351, pp. 329–337. Springer, Heidelberg (1997). doi:10.1007/3-540-63930-6_138
25. Ferri, M., Gallina, S., Porcellini, E., Serena, M.: On-line character and writer recognition by size functions and fuzzy logic. In: Proceedings of ACCV 1995, pp. 5–8 (1995)
26. Ferri, M., Lombardini, S., Pallotti, C.: Leukocyte classifications by size functions. In: Proceedings of the Second IEEE Workshop on Applications of Computer Vision, pp. 223–229. IEEE (1994)
27. Ferri, M., Stanganelli, I.: Size functions for the morphological analysis of melanocytic lesions. J. Biomed. Imaging **2010**, 5 (2010)
28. Ferri, M., Tomba, I., Visotti, A., Stanganelli, I.: A feasibility study for a persistent homology-based k-nearest neighbor search algorithm in melanoma detection. J. Math. Imaging Vis. **57**, 1–16 (2016)
29. Frosini, P., Jabłoński, G.: Combining persistent homology and invariance groups for shape comparison. Discrete Comput. Geom. **55**(2), 373–409 (2016)
30. Giorgi, D., Frosini, P., Spagnuolo, M., Falcidieno, B.: 3D relevance feedback via multilevel relevance judgements. Vis. Comput. **26**(10), 1321–1338 (2010)
31. Giusti, C., Pastalkova, E., Curto, C., Itskov, V.: Clique topology reveals intrinsic geometric structure in neural correlations. Proc. Natl. Acad. Sci. **112**(44), 13455–13460 (2015)
32. Handouyahia, M., Ziou, D., Wang, S.: Sign language recognition using moment-based size functions. In: Proceedings of International Conference on Vision, Interface, pp. 210–216 (1999)
33. Hatcher, A.: Algebraic Topology. Cambridge University Press, New York (2001)
34. Holzinger, A.: On knowledge discovery and interactive intelligent visualization of biomedical data. In: Proceedings of the International Conference on Data Technologies and Applications DATA 2012, Rome, Italy, pp. 5–16 (2012)
35. Holzinger, A.: On topological data mining. In: Holzinger, A., Jurisica, I. (eds.) Interactive Knowledge Discovery and Data Mining in Biomedical Informatics. LNCS, vol. 8401, pp. 331–356. Springer, Heidelberg (2014). doi:10.1007/978-3-662-43968-5_19
36. Kelly, D., McDonald, J., Lysaght, T., Markham, C.: Analysis of sign language gestures using size functions and principal component analysis. In: Machine Vision and Image Processing Conference, IMVIP 2008. International, pp. 31–36. IEEE (2008)
37. Lamar-León, J., García-Reyes, E.B., Gonzalez-Diaz, R.: Human gait identification using persistent homology. In: Alvarez, L., Mejail, M., Gomez, L., Jacobo, J. (eds.) CIARP 2012. LNCS, vol. 7441, pp. 244–251. Springer, Heidelberg (2012). doi:10.1007/978-3-642-33275-3_30

38. Liu, J.-Y., Jeng, S.-K., Yang, Y.-H.: Applying topological persistence in convolutional neural network for music audio signals. arXiv preprint arXiv:1608.07373 (2016)

39. Lord, L.-D., Expert, P., Fernandes, H.M., Petri, G., Van Hartevelt, T.J., Vaccarino, F., Deco, G., Turkheimer, F., Kringelbach, M.L.: Insights into brain architectures from the homological scaffolds of functional connectivity networks. Front. Syst. Neurosci. **10**, 85 (2016)

40. Micheletti, A.: The theory of size functions applied to problems of statistical shape analysis. In: S4G-International Conference in Stereology, Spatial Statistics and Stochastic Geometry, pp. 177–183. Union of Czech Mathematicians and Physicists (2006)

41. Micheletti, A., Landini, G.: Size functions applied to the statistical shape analysis and classification of tumor cells. In: Bonilla, L.L., Moscoso, M., Platero, G., Vega, J.M. (eds.) ECMI 2006. Mathematics in Industry, vol. 12, pp. 538–542. Springer, Heidelberg (2008)

42. Petri, G., Expert, P., Turkheimer, F., Carhart-Harris, R., Nutt, D., Hellyer, P.J., Vaccarino, F.: Homological scaffolds of brain functional networks. J. Roy. Soc. Interface **11**(101), 20140873 (2014)

43. Platt, D.E., Basu, S., Zalloua, P.A., Parida, L.: Characterizing redescriptions using persistent homology to isolate genetic pathways contributing to pathogenesis. BMC Syst. Biol. **10**(1), S10 (2016)

44. Port, A., Gheorghita, I., Guth, D., Clark, J.M., Liang, C., Dasu, S., Marcolli, M.: Persistent topology of syntax. arXiv preprint arXiv:1507.05134 (2015)

45. Ringermacher, H.I., Mead, L.R.: A new formula describing the scaffold structure of spiral galaxies. Mon. Not. R. Astron. Soc. **397**(1), 164–171 (2009)

46. Sizemore, A., Giusti, C., Betzel, R.F., Bassett, D.S.: Closures and cavities in the human connectome. arXiv preprint arXiv:1608.03520 (2016)

47. Stanganelli, I., Brucale, A., Calori, L., Gori, R., Lovato, A., Magi, S., Kopf, B., Bacchilega, R., Rapisarda, V., Testori, A., Ascierto, P.A., Simeone, E., Ferri, M.: Computer-aided diagnosis of melanocytic lesions. Anticancer Res. **25**(6C), 4577–4582 (2005)

48. Turner, K., Mukherjee, S., Boyer, D.M.: Persistent homology transform for modeling shapes and surfaces. Inf. Inference J. IMA **3**(4), 310–344 (2014)

49. Uras, C., Verri, A.: On the recognition of the alphabet of the sign language through size functions. In: Proceedings of the 12th IAPR International Conference on Pattern Recognition, Vol. 2-Conference B: Computer Vision & Image Processing, vol. 2, pp. 334–338. IEEE (1994)

50. Verri, A., Uras, C., Frosini, P., Ferri, M.: On the use of size functions for shape analysis. Biol. Cybern. **70**, 99–107 (1993)

51. Ziefle, M., Himmel, S., Holzinger, A.: How usage context shapes evaluation and adoption criteria in different technologies. In: AHFE 2012, Proceeding of International Conference on Applied Human Factors and Ergonomics, San Francisco, pp. 2812–2821 (2012)

Predictive Models for Differentiation Between Normal and Abnormal EEG Through Cross-Correlation and Machine Learning Techniques

Jefferson Tales Oliva[✉] and João Luís Garcia Rosa

Bioinspired Computing Laboratory, Institute of Mathematics and Computer Science,
University of São Paulo, São Carlos, São Paulo 13566–590, Brazil
jeffersonoliva@usp.br, joaoluis@icmc.usp.br

Abstract. Currently, in hospitals and medical clinics, large amounts of data are becoming increasingly registered, which usually are derived from clinical examinations and procedures. An example of stored data is the electroencephalogram (EEG), which is of high importance to the various diseases that affect the brain. These data are stored to keep the patient's clinical history and to help medical experts in performing future procedures, such as pattern discovery from specific diseases. However, the increase in medical data storage makes unfeasible their manual analysis. Also, the EEG can contain patterns difficult to be observed by naked eye. In this work, a cross-correlation technique was applied for feature extraction of a set of 200 EEG segments. Afterwards, predictive models were built using machine learning algorithms such as J48, 1NN, and BP-MLP (backpropagation based on multilayer perceptron), that implement decision tree, nearest neighbor, and artificial neural network, respectively. The models were evaluated using 10-fold cross-validation and contingency table methods. The evaluation results showed that the model built with the J48 performed better and was more likely to correctly classify EEG segments in this study than 1NN and BP-MLP, corresponding to 98.50% accuracy.

Keywords: EEG · Cross-correlation · Machine learning · Predictive models · Classification

1 Introduction

Technological advances over the years have increased capability of data processing and storage [15]. In this sense, large amounts of information are increasingly stored in databases. In medical area, data are generated from different sources and represented in various formats such as text, video, audio, and image [24,33]. These data are often obtained by medical examinations, as the electroencephalography, whose records are called electroencephalograms (EEG), which is the monitoring result over time of the variation of electrical activity

A. Holzinger et al. (Eds.): Integrative Machine Learning, LNAI 10344, pp. 134–145, 2017.
https://doi.org/10.1007/978-3-319-69775-8_7

generated by synapses among neuron populations and it is highly important for the diagnosis of many brain diseases [4, 11]. The EEG can be represented as time series (TS), which are sets of observations ordered in time [5].

Also, medical examinations are stored in databases in order to maintain the patients' clinical history and be reused in health domain in decision making processes for diseases diagnosis and for conducting future procedures, *e.g.*, for collecting tissue samples [24].

However, with the increasing storage of information in medical databases, their manual analysis becomes an infeasible task. Also, the EEG may contain standards that are difficult to observe by naked eye, being necessary the development of methods and tools to assist in the analysis and management of these data [13].

According the World Health Organization (WHO)[1], mental and neurological diseases affect approximately 700 million people in the world, of which one-third do not have medical monitoring. Also, the neurological disorders will result in the loss of 16.3 million American dollars between 2011 and 2030 [35].

In this sense, many computational methods have been developed and applied in different fields to assist in data analysis and management, data mining (DM) process supported by machine learning (ML) methods, which have attracted the interest of several researchers for the building of descriptive and predictive models from implicit knowledge existing in the data [34]. For the application of these techniques, the data should be represented in an appropriate format, such as attribute-value table. To do so, several features can be extracted from EEG signals and they need to be explored [16].

Thus, in this work, predictive models were built using DM supported by ML techniques, such as decision tree, 1-nearest-neighbor and artificial neural network, in order to classify EEG segments into normal or abnormal class.

This paper is organized as follows: Sect. 2 presents a glossary and key terms related to this paper; Sect. 3 describes the database used in this work, the method used to extract features in EEG and the techniques used to build and evaluate the predictive models; Sect. 4 presents the results and discussion in terms of the classification efficiency achieved by applying the proposal in the database; Sect. 5 reports the final highlights; and Sect. 6 describes proposals for future works.

2   Glossary and Key Terms

Epilepsy is a neurological disorder characterized by seizures [21].

Epileptic seizures are brief occurrences of signals and/or symptoms as a result of disturbances in the electrical activity of the brain [9].

International 10–20 system is a method for distribution of electrodes on the scalp in order to collect EEG signals. In this system, the electrodes are divided into specific locations, considering a total distance around between 10 and 20% of the head circumference [11].

---

[1] http://www.who.int.

Peak value is the maximum value of a time series (TS).

Instant value is the wave value at any particular instant related with peak value. The instant value can be calculated by multiplying the peak value by a quarter of a sine wave ($sin(45°)$ or 0.707) [19].

Centroid is the geometric center of a TS [8].

Equivalent width is the wave width from the peak value of a TS [8];

Mean square abscissa is the spreading of TS amplitude on the centroid [8].

3   Materials and Methods

3.1   EEG Dataset

A public EEG database was used in this work [2]. The EEG signals were generated by a 128-channel amplifier system, that used an average common reference. These signals were generated at a sampling rate of 173.61 Hz using 12-bit resolution and filtered by band-pass at 0.53–40 Hz (12 dB/oct.). The electrode placement used was the international 10–20 system.

In this database there are 100 single channel EEG segments with duration of 23.6 s obtained from different subjects in each of five different sets. These segments were selected and artifacts, such as muscle activity or eye movements, were removed. The recording conditions of each set are the following:

- **A:** recordings of healthy volunteers with eyes open;
- **B:** recordings of healthy volunteers with eyes closed;
- **C:** recordings of the hippocampal formation of the opposite hemisphere of the brain from patients with epilepsy;
- **D:** recordings of the epileptogenic zone from patients with epilepsy;
- **E:** recordings of the seizure activity, selected from all recording sites exhibiting ictal activity from patients with epilepsy.

In this work, only two sets are considered, set A (normal) and E (abnormal), amounting to 200 EEG segments, according with previous works [3,17]. Figure 1 shows a health EEG example. Figure 2 shows an epileptic EEG example.

3.2   Feature Extraction

Feature extraction consists in an essential task for representation of EEG signals and it influences the classification performance [17]. In this sense, features can be extracted based in a mathematical operation named cross-correlation (CC) [3], which measures the extent of similarity between two signals [25]. The CC of the signals x and y can be given by Eq. 1, where n is the signal size and m represents the time shift parameter and it is denoted by $m = \{..., -3, -2, -1, 0, 1, 2, 3, ...\}$.

$$CC(x, y, m) = \begin{cases} \sum_{i=0}^{n-m-1} x_{i+m} * y_i & m \geq 0 \\ CC(y, x, -m) & m < 0 \end{cases} \tag{1}$$

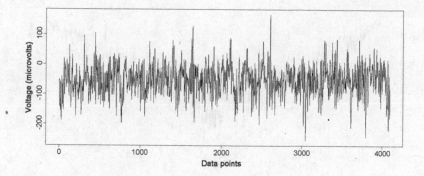

Fig. 1. Normal EEG segment sample.

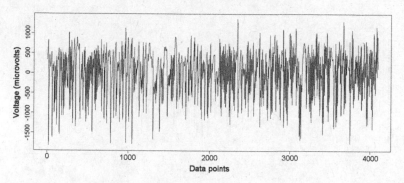

Fig. 2. Epileptic EEG segment sample.

For two TS with length n, the CC with m values between $-n$ and n is measured, generating a cross-correlogram (CCo) with length $2 * n - 1$, *i.e.*, the j-*th* cross-correlogram value is the CC measured for $m = j - n$.

Figure 3 shows the cross-correlogram of two healthy EEG segments, Fig. 4 shows the cross-correlogram of an epileptic and healthy EEG segments and Fig. 5 shows the cross-correlogram of two epileptic EEG segments.

The following features can be extracted from the cross-correlogram [3]:

- **Peak value (F1):**

$$F1 = max(CCo) \qquad (2)$$

- **Instant value (F2):**

$$F2 = 0.707 * F1 \qquad (3)$$

- **Centroid (F3):**

$$F3 = \frac{\sum_{i=-n}^{n} i * CCo(i)}{\sum_{i=-n}^{n} CCo(i)} \qquad (4)$$

138   J.T. Oliva and J.L. Garcia Rosa

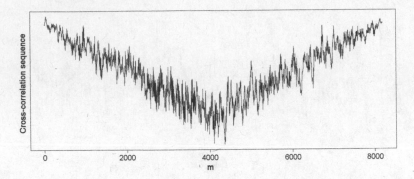

Fig. 3. Cross-correlogram of two healthy EEG segments.

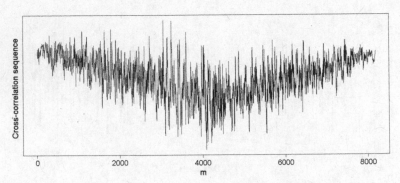

Fig. 4. Cross-correlogram of an epileptic and healthy EEG segments.

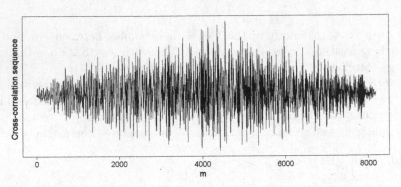

Fig. 5. Cross-correlogram of two epileptic EEG segments.

- **Equivalent width (F4):**

$$F4 = \frac{\sum_{i=-n}^{n} CCo(i)}{F1} \tag{5}$$

- **Mean square abscissa (F5):**

$$F5 = \frac{\sum_{i=-n}^{n} i^2 * CCo(i)}{\sum_{i=-n}^{n} CCo(i)} \qquad (6)$$

In this sense, in the k-th EEG segment these features are extracted, generating the instance $I_k = \{F1_k, F2_k, F3_k, F4_k, F5_k, C_k\}$, where C is the class of the k-th EEG. This instance format is required for the classifiers induction by ML algorithms. For feature extraction using CCo, firstly, an EEG segment is selected as unique reference, decreasing by 1 the number of instances. Next, this reference is used for building CCo of all other EEG segments [17].

3.3 Building of Classification Models

After feature extraction, predictive models were built using ML techniques: (1) decision tree (DT), (2) nearest-neighbor (NN), and (3) artificial neural network (ANN).

Method (1) builds a model, in which its data structure is hierarchically organized. In this structure, different classes are represented by a set of rules, that are derived from the tree. To classify new examples, the tree is traversed from its root to verify values of the features and define the class of the analyzed example [26, 29].

Method (2) is a classification technique that classifies a new example by calculating its similarity with the training set examples, which have been previously labeled by experts. This similarity can be measured using distance measures, such as the Euclidean distance. The NN method does not build a predictive model and it is based only in memory, *i.e.*, the model is the training set itself [1].

Method (3) builds mathematical models inspired in the biological neural structure, which have the computational capacity acquired by means of learning and generalization. The most common algorithm used for ANN training are based in error correction [14].

3.4 Model Evaluation

The built models are evaluated according to its predictive quality, considering the efficiency for classification of new examples. This evaluation can be done using cross-validation (CV) [22] and contingency table (CT) [10] methods.

The CV method divides the data set into k equal samples, which the k-th sample is the test set and the k-1 remaining samples correspond to the training set. Thus, each element of the test is classified by the built model through corresponding training set. Following, the average error and standard deviation measures regarding the classification performance of the k samples are calculated [22]. Also, to complement the evaluation of the results, statistical tests can be performed to compare the classification models in order to verify the existence of the statistically significant difference between two or more models, considering a significance level.

The CTs are used to evaluate relationships between two or more nominal variables, *i.e.*, if it belongs or doesn't belong to the same class. In CT, the following accuracy attributes can be extracted [10]:

- **Positive predictive value (PPV):** is the percentage of abnormal examples in relation to the total of instances that were classified into the abnormal class;
- **Negative predictive value (NPV):** is the percentage of normal examples in relation to the total of instances that were classified into the normal class;
- **Sensitivity:** is the percentage of examples that were classified into the abnormal class related to the total number of instances in the set of examples;
- **Specificity:** is the percentage of examples that were classified into the normal class related to the total number of instances in the set of examples.

For the feature extraction, a tool was implemented in Java[2] language using a software development platform named NetBeans[3]. The building and evaluation of the prediction models were performed using the WEKA tool [20], which contains ML algorithms for building models, such as: J48, an implementation of the C4.5 algorithm [27], that is used for induction of DT; 1NN (1-nearest-neighbor) [1] for classification of examples by calculating their similarity with the training set examples; and BP-MLP (backpropagation based on multilayer perceptron) [14] for building ANN. The statistical analysis can be performed using the software GraphPad Instat©.

4   Results and Discussion

Other works employed similar approaches. In [7], CCo with support vector machines (SVM) was applied in order to classify ECG beats. In [28], CCo with SVM was used for classification of motor imagery (MI). In [31], CCo with logistic regression techniques [32] was used in order to identify MI tasks. In [30], CCo and the artificial divergent autoencoder neural network [18] were applied in order to test the feasibility of EEG authentication. Also, CCo with ML techniques was used in other related works [3,12,23,36]. The accurate results shown in such studies were confirmed by the results obtained here.

Although these studies have shown good results, their evaluation was based on accuracy, or other measures based on CT. Accordingly, the application of significance tests (statistical tests) is required to augment the evaluation of classifiers in order to verify the existence of the statistically significant difference among these models.

In this work, classification models were built using the J48, 1NN, and BP-MLP algorithms and evaluated using the CV and CT methods. This evaluation was complemented by statistical tests. Also, it is important to emphasize that, unlike other work, this paper compares the performance among the J48, 1NN

---

[2] http://www.oracle.com/technetwork/java/index.html.
[3] https://netbeans.org/.

and BP-MLP algorithms for the differentiation of epileptic and healthy EEG segments.

Thus, in an experimental evaluation, features based on CCo were extracted in a set of 200 EEG segments, being 100 for normal EEG (set A) and 100 for abnormal (or epileptic) EEG (set E). Subsequently, the first abnormal EEG of a set was selected as reference for the CCo building and its respective feature extraction. Each EEG segment was represented by a set of five features.

Afterwards, ML techniques were applied to build predictive models using J48 (DT), 1NN (NN), and BP-MLP (ANN) algorithms by WEKA tool to induce models.

Posteriorly, the built models from the previous step were evaluated based on predictive accuracy by analyzing the results obtained by CV and CT methods.

The CV method was applied taking ten partitions (10-fold) into account. From these partitions, the mean error (ME) and the standard deviation (SD) measures were calculated for each model evaluated. Table 1 summarizes the evaluation results based on average error and respective standard deviation.

Table 1. Results of applying the CV method to evaluate the predictive models.

Algorithm	Average error (%)	Standard deviation (%)
J48	1.500	2.415
1NN	8.079	5.493
BP-MLP	11.710	11.674

Based on this table, it was found that the model built with the J48 algorithm presents smaller average error and SD than the 1NN and BP-MLP. In order to verify the occurrence of a statistically significant difference between the models, a statistical test was performed for paired data. For selecting the appropriate test type, the p-value normality test was applied in the error values generated by 10-fold CV for each model. With the application of this normality test, only the model generated by J48 algorithm wasn't approved, determining that the test to be applied should be non parametric. Thus, the Friedman test was applied, considering the significance level of 5% [10], leading to very significant difference with a p-value of 0.0029. To identify the model in which a very significant difference occurred, the Dunn's post hoc test [6] was applied, which states that there was a statistically significant difference (p-value > 0.05) only between 1NN and BP-MLP. Thus, according to the statistical tests used in this study, the model built with J48 algorithm obtained better performance than the other models.

Another technique used to evaluate the models was the CT. Tables 2, 3 and 4 show the correspondent results.

According to these tables, the model built using the J48 algorithm correctly classified 99 EEG segments without abnormalities and 97 with abnormalities, thus yielding better performance than the one built by the 1NN and BP-MLP algorithms.

Table 2. CT for the model built by the J48 algorithm.

Classification	Normal	Abnormal	Total
Normal	99	1	100
Abnormal	2	97	99
Total	101	98	199

Table 3. CT for the model built by the 1NN algorithm.

Classification	Normal	Abnormal	Total
Normal	93	7	100
Abnormal	9	90	99
Total	102	97	199

Table 4. CT for the model built by the BP-MLP algorithm.

Classification	Normal	Abnormal	Total
Normal	90	10	100
Abnormal	13	86	99
Total	103	96	199

Table 5. Measures calculated in each contingency table.

Algorithm	PPV	NPV	Sensitivity	Specificity
J48	98.02%	98.98%	99.00%	97.98%
1NN	91.96%	92.78%	93.00%	90.91%
BP-MLP	87.38%	89.58%	90.00%	86.87%

Table 5 presents four precision measures for each CT built.

Table 5 shows that the model built using the J48 algorithm obtained the highest values for the parameters PPV, NPV, sensitivity, and specificity, which were measured as 98.02%, 98.98%, 99.00% and 97.98%, respectively. Therefore, the model using the J48 algorithm performed better and was more likely to correctly classify EEG segments than the 1NN and BP-MLP algorithms.

5   Conclusion

This paper compares the performance among the J48, 1NN, and BP-MLP algorithms for the differentiation of epileptic and healthy EEG segments. For this, an approach to feature extraction in EEG segments and to build predictive models using CCo and ML techniques, respectively, was applied.

The CCo and its features were extract by a tool implemented in Java language. In this sense, a CCo was extracted for each EEG through its correlation with the EEG segment selected as reference. The predictive models were built using WEKA tool, that contains ML algorithms as J48, 1NN, and BP-MLP. For evaluating the models we used 10-fold CV and CT methods.

Thus, the experimental evaluation was conducted in a set of EEG segments, that is divided in five subsets, according to the description in Sect. 3.1. In this evaluation, the predictive models were applied to classify each EEG segment (represented by CCo) into normal or abnormal class.

The performance evaluation of the models using CV found statistically very significant difference (p-value of 0.0029) between them. After performing Friedman test, Dunn's post hoc test was applied to identify the model in which the significant difference occurred, evidencing that the model built with the J48 performed better and was more likely to correctly classify EEG segments in this study than 1NN and BP-MLP.

The evaluation using CT found that the model built with the J48 algorithm performed better and was more likely to correctly classify EEG segments than the other models used in this work, corroborating the previous evaluation.

6 Future Research

Future works include performing feature extraction by CCo using other EEG segments as reference, studying and implementing other feature extraction techniques in order to expand and improve EEG representation, studying real EEG sets related to epilepsy and other diseases diagnosed by EEG, building more representative predictive models, using others ML techniques for building models, and using CCo technique in EEG processing for automatic generation of medical reports.

So, it is expected the building of more accurate classifiers, a greater capacity for epileptic seizure prediction, the classification of EEG signals related to different neural diseases, the construction of a tool for filling in medical reports automatically, and the assistance in the decision making processes by medical experts.

Acknowledgment. I would like to thank the Brazilian funding agency Coordenação de Aperfeiçoamento de Pessoal de Nível Superior (CAPES) for financial support.

References

1. Alpaydin, E.: Introduction to Machine Learning. MIT Press, Cambridge (2014)
2. Andrzejak, R.G., Lehnertz, K., Mormann, F., Rieke, C., David, P., Elger, C.E.: Indications of nonlinear deterministic and finite-dimensional structures in time series of brain electrical activity: dependence on recording region and brain state. Phys. Rev. E **64**(6), 061907 (2001)
3. Chandaka, S., Chatterjee, A., Munshi, S.: Cross-correlation aided support vector machine classifier for classification of EEG signals. Expert Syst. Appl. **36**(2), 1329–1336 (2009)

4. Chaovalitwongse, W.A., Prokopyev, O.A., Pardalos, P.M.: Electroencephalogram (EEG) time series classification: applications in epilepsy. Ann. Oper. Res. **148**(1), 227–250 (2006)
5. Chiu, B., Keogh, E., Lonardi, S.: Probabilistic discovery of time series motifs. In: Proceedings of the International Conference on Knowledge Discovery and Data Mining, Washington, USA, pp. 493–498 (2003)
6. Dunn, O.J.: Multiple comparisons using rank sums. Technometrics **6**(3), 241–252 (1964)
7. Dutta, S., Chatterjee, A., Munshi, S.: Correlation technique and least square support vector machine combine for frequency domain based ECG beat classification. Med. Eng. Phys. **32**(10), 1161–1169 (2010)
8. Easton Jr., R.L.: Fourier Methods in Imaging. Wiley, Danvers (2010)
9. Fisher, R.S., Boas, W.E., Blume, W., Elger, C., Genton, P., Lee, P., Engel, J.: Epileptic seizures and epilepsy: definitions proposed by the international league against epilepsy (ILAE) and the international bureau for epilepsy (IBE). Epilepsia **46**(4), 470–472 (2005)
10. Fredman, D., Pisani, R., Ourvers, R.: Statistics. Norton, New York (1988)
11. Freeman, W.J., Quian Quiroga, R.: Imaging Brain Function with EEG: Advanced Temporal and Spatial Analysis of Electroencephalographic Signals. Springer, New York (2013)
12. Gajic, D., Djurovic, Z., Gligorijevic, J., Di Gennaro, S., Savic-Gajic, I.: Detection of epileptiform activity in EEG signals based on time-frequency and non-linear analysis. Front. Comput. Neurosci. **9**(38), 1–16 (2015)
13. Han, J.: Data Mining: Concepts and Techniques. Morgan Kaufmann Publishers, San Francisco (2006)
14. Haykin, S.: Neural Networks and Learning Machines. Pearson Education, Upper Saddle River (2009)
15. Hilbert, M., López, P.: The world's technological capacity to store, communicate, and compute information. Sci. Mag. **332**(6025), 60–65 (2011)
16. Holzinger, A., Scherer, R., Seeber, M., Wagner, J., Müller-Putz, G.: Computational sensemaking on examples of knowledge discovery from neuroscience data: towards enhancing stroke rehabilitation. In: Böhm, C., Khuri, S., Lhotská, L., Renda, M.E. (eds.) ITBAM 2012. LNCS, vol. 7451, pp. 166–168. Springer, Heidelberg (2012). doi:10.1007/978-3-642-32395-9_13
17. Iscan, Z., Dokur, Z., Demiralp, T.: Classification of electroencephalogram signals with combined time and frequency features. Expert Syst. Appl. **38**(8), 10499–10505 (2011)
18. Kurtz, K.J.: The divergent autoencoder (DIVA) model of category learning. Psychon. Bull. Rev. **14**(4), 560–576 (2007)
19. Learn about Electronics: Measuring the sine wave (2015). http://www.learnabout-electronics.org/ac_theory/ac_waves02.php
20. Machine Learning Group, The Universisty of Waikato: Weka 3: Data mining software in Java (2015). http://www.cs.waikato.ac.nz/ml/weka/
21. Magiorkinis, E., Sidiropoulou, K., Diamantis, A.: Hallmarks in the history of epilepsy: epilepsy in antiquity. Epilepsy Behav. **17**(1), 103–108 (2010)
22. McLachlan, G., Do, K., Ambroise, C.: Analyzing Microarray Gene Expression Data. Wiley, Danvers (2005)
23. Nataraj, S.K., bin Yaacob, S., Paulraj, M.P., Adom, A.H.: EEG based intelligent robot chair with communication aid using statistical cross correlation based features. In: Proceedings of the International Conference on Bioinformatics and Biomedicine, Belfast, UK, pp. 12–18 (2014)

24. Oliva, J.T.: Automating the process of mapping medical reports to estructured database. Master thesis, State University of West Paraná, Foz do Iguaçu, Brazil (2014)
25. Proakis, J.G., Manolakis, D.K.: Digital Signal Processing: Principles, Algorithms, and Application. Prentice Hall, Saddle River (2006)
26. Quinlan, J.R.: Simplifying decision trees. Int. J. Man Mach. Stud. **27**(3), 221–234 (1987)
27. Quinlan, J.R.: C4.5: Programs for Machine Learning. Elsevier, San Francisco (2014)
28. Rathipriya, N., Deepajothi, S., Rajendran, T.: Classification of motor imagery ECoG signals using support vector machine for brain computer interface. In: Proceedings of the International Conference Advanced Computing, Chennai, India, pp. 63–66 (2013)
29. Rezende, S.O.: Sistemas Inteligentes: Fundamentos e Aplicações. Manole, Barueri (2003). (in Portuguese)
30. Ruiz Blondet, M.V., Khalifian, N., Kurtz, K.J., Laszlo, S., Jin, Z.: Brainwaves as authentication method: proving feasibility under two different approaches. In: Proceedings of the 40th Northeast Bioengineering Conference, Boston, USA, pp. 1–2 (2014)
31. Siuly, S., Li, Y., Wen, P.: Identification of motor imagery tasks through CC-LR algorithm in brain computer interface. Int. J. Bioinform. Res. Appl. **9**(2), 156–172 (2013)
32. Walker, S.H., Duncan, D.B.: Estimation of the probability of an event as a function of several independent variables. Biometrika **54**(1–2), 167–179 (1967)
33. Wiederhold, G., Shortliffe, E.H.: System design and engineering in health care. In: Shortliffe, E.H., Cimino, J.J. (eds.) Biomedical Informatics: Computer Applications in Health Care and Biomedicine. Health Informatics, pp. 233–264. Springer, New York (2006). doi:10.1007/0-387-36278-9_6
34. Witten, I., Frank, E., Hall, M.A.: Machine Learning: Practical Machine Learning Tools and Techniques. Morgan Kaufmann, San Francisco (2011)
35. World Health Organization: Draft comprehensive mental health action plan 2013–2020 (2015). http://apps.who.int/gb/ebwha/pdf_files/EB132/B132_8-en.pdf
36. Xielifuguli, K., Fujisawa, A., Kusumoto, Y., Matsumoto, K., Kita, K.: Pleasant/unpleasant filtering for affective image retrieval based on cross-correlation of EEG features. Appl. Comput. Intell. Soft Comput. **2014** (2014)

A Brief Philosophical Note on Information

Vincenzo Manca[(⊠)]

Departement of Computer Science, Center of BioMedical Computation,
University of Verona, Verona, Italy
vincenzo.manca@univr.it

1 Introduction

I will start by posing a question that arose to my attention when, some years ago, I realized the importance of Machine Learning for the future theoretical and applicative fields of Computer science [10,12–14]. Why Machine Learning (ML) is so attractive and popular, so versatile in the applications and so powerful in providing solutions? ML is not the main subject of my research, but continually I meet tools and approaches that refer essentially to the field of Machine Learning [3,15–17]. In this short note I will try to express a possible answer to the question and some comments related to it.

Of course ML is, first at all, as any Computer Science field, related to information. But this alone is not an explanation of its success. A crucial ingredient, peculiar to ML, is its natural bivalent nature connecting two crucial and deep aspects of Information: probabilistic and digital information.

2 Prologue

In his epochal booklet [21], Shannon founded Information Theory by considering the basic notion of Information Source (IS), which is essentially a discrete distribution of probability. The quantity of information of an event or of a datum is a function of its probability, with respect to a given distribution. An information measure has to be a function that increases when the probability decreases (rare events are informative) and such that is additive for two events that are independent: $Inf(E_1, E_2) = Inf(E_1) + Inf(E_2)$ if $P(E_1, E_2) = P(E_1)P(E_2)$ [1,8,9,11]. The simplest mathematical function having these properties is $-\lg$, the minus logarithm (the chosen base determines a multiplicative factor). Therefore, $Inf(E) = -lg(P(E))$. A direct consequence of this definition is that the entropy of an Information source corresponds to the probabilistic mean of the information quantities of data emitted by the information source [8,21]. This short account of Shannon approach wants to stress the novelty of his point of view. Information is essentially probability (or statistics when probabilities are given as frequencies of observed data, the relationship between probability and statistics is more complex and subtle, but here we cannot enter in further details [22]).

© Springer International Publishing AG 2017
A. Holzinger et al. (Eds.): Integrative Machine Learning, LNAI 10344, pp. 146–149, 2017.
https://doi.org/10.1007/978-3-319-69775-8_8

The novelty of Shannon probabilistic viewpoint is very strong, because the trend of the early theoretical speculations in Computer Science, developed since the epochal Turing paper [23], were based on the notion of symbolic sequence. The amount of information of a symbolic sequence is essentially its length. This length corresponds to the digital information of the encoded datum, which of course, depends on the encoding method. As Shannon proved, and the further developments of Information Theory confirmed, probabilistic and digital information are strongly related and very often they express complementary aspects of the same reality, which without their integration could result inaccessible or hard to be rigorously explicated [8]. The first two Shannon theorems (the third one, related to signals, is of continuous/analogical nature) are two jewels in proving as fruitful is the integration of probabilistic and digital aspects. In fact, First Theorem claims, under very general hypotheses (of unique decipherable codes), that the average length of a code for an information source $S = (A, p)$ (A are the data and p their probabilities) is lower bounded by the entropy $H(A, p) = -\sum_{x \in A} p(x) \lg p(x)$. This means that no code can reduce the average length of its codewords under the limit stated by the entropy. Second Theorem is more difficult to express in a succinct way, but it essentially proves that in a transmission process we can reduce the probability of error to zero (error means receiving something different from the message that was sent). To this end, it is enough to transmit by using more symbols of those that are necessary to encode all the messages (for example, strings of length n over two symbols, in such a way that $2^n > m$, where m is the number of possible messages). The extra symbols, usually called control or parity symbols are employed by self correcting codes (discovered after Shannon theorem) for recognizing errors, and possibly, for recovering the right symbols. These codes apply algebraic and algorithmic methods, but the proof given by Shannon was entirely probabilistic (based on typical sequences, conditional entropy and mutual information).

3   Epilogue

The scientific strength of ML resides in continuing the spirit underlying Shannon's booklet, plus the computational power of software technology, and plus the use of inverse methods [16]. However, what it seems to be essential and peculiar to ML is its statistical basis coupled with the algorithmic and computer elaboration of digital information of data.

Here, I want to express a challenge that again goes back to Shannon's booklet and is surely of interest for developing theories and applications of ML. The Sect. 6 of [21] is entitled *Choice, Uncertainty and Entropy* and it begins with the question: "Can we define a quantity which will measure, in some sense, how much information is "produced" by such a process? [an information source]". The section states some conditions to which this measure has to verify, claiming that entropy H is a good measure of information (recalling Boltzmann's H function and the H Theorem [2,5,7,18–20]). The section concludes by asserting that $H(A, p)$ corresponds to the *uncertainty* associated to the information source

(A, p). I was always impressed by this passage of the book. Shannon does not intend to enter into philosophical details [4, 6, 25]. He wants to prove theorems mainly useful in information transmission. However, it is almost sure that he is conscious of the apparent paradox of identifying uncertainty with information. Il would be as identifying ignorance with knowledge. Surely, this paradox is related with the probabilistic notion of event. The probability of an event is a measure of the (*a priori*) uncertainty before the event happens, but it is also a measure of the (*a posteriori*) information after it happens. However, I think that the problem is more complex and is a clue about extensions of the notion of information source. Let us consider a relational viewpoint in considering information. The quantity of information of S with respect to O is proportional to the reduction of the event space of S when S receives an amount of information from O "observing" events of the event space of O. This resembles Shannon's notion of mutual information that he obtains as a derived notion. In more formal terms, at any time t, for both S and O, two sets of possible events $E_t(S)$ and $E_t(O)$ are defined and an information interaction $O \rightarrow S$ makes $E_{t+1}(S) \neq E_t(S)$. This means that information depends on the kind of interactions that are considered and on the verse of its flow. In fact, the information of O with respect to S can be, in general, different from the information of S with respect to O. Let us consider a system O of molecules of an ideal gas within a given volume without interactions with the environment. Molecules collide and in each collision a molecule determines position and speed of the molecules with which it collides. On the average, the internal information exchanged among gas molecules increases (corresponding to the increasing of physical entropy). But, let us assume that a very improbable difference of pressure between two opposite sides of gas volume can be recognized by some sensor S activating a particular process. In this perspective, other event spaces have to be considered, and the amount of information passing from O to S is greater if the internal information of O decreases. Therefore, an information source intended as an absolute notion does not capture important aspects such as the verse and the level at which information is exchanged (gas could be less informative for S when the maximum of its entropy is reached, that is, when the gas has the maximum value of average (internal) information).

In conclusion, according to a relational perspective, an IS has to be considered with respect to another IS when its probability distribution E_t changes in time by interacting with another IS. In fact, it is apparent in many situations that not only information is determined by probability, but it can also change probability spaces, according to the Bayesian perspective in probability. At same time, uncertainty can be viewed as a powerful way of managing with all the states belonging to an uncertain or indeterminate state (was Shannon aware of this when he entitled Sect. 6?). This possibility is the main point of quantum computing [24]. In conclusion, extending the notion of information source in relational terms could open new possibilities for next generation ML, which not only continues Shannon perspective, but moves it toward new frontiers of investigations and applications [24].

References

1. Aczel, A.D.: Chance. Thunder's Mouth Press, New York (2004)
2. Boltzmann, L.: Weitere Studien uber das Wärmegleichgewicht unter Gasmolekulen, Sitzungsber. Kais. Akad. Wiss. Wien Math. Naturwiss. Classe 66, 275–370 (1872)
3. Bonnici, V., Manca, V.: Informational laws of genome structures. Scientific Reports 6, 28840 (2016). http://www.nature.com/articles/srep28840
4. Brillouin, L.: The negentropy principle of information. J. Appl. Phys. **24**, 1152–1163 (1953)
5. Brush, S.G., Hall, N.S. (eds.): The Kinetic Theory of Gases: An Anthology of Classical Papers with Historical Commentary. Imperial College Press, London (2003)
6. Calude, C.S.: Information and Randomness: An Algorithmic Perspective. EATCS Series in Theoretical Computer Science. Springer, Heidelberg (1994)
7. Carnot, S.: Reflections on the Motive Power of Heat (English translation from French edition of 1824, with introduction by Lord Kelvin). Wiley, New York (1890)
8. Cover, T., Thomas, C.: Information Theory. Wiley, New York (1991)
9. Feller, W.: An Introduction to Probability Theory and Its Applications. Wiley, New York (1968)
10. James, G., Witten, D., Hastie, T., Tibshirani, R.: An Introduction to Statistical Learning. Springer, New York (2013)
11. Jaynes, E.T.: Information Theory and Statistical Mechanics. Phys. Rev. **33**(5), 620–630 (1957)
12. Holzinger, A., Jurisica, I. (eds.): Interactive Knowledge Discovery and Data Mining in Biomedical Informatics. LNCS, vol. 8401. Springer, Heidelberg (2014)
13. Holzinger, A., Hortenhuber, M., Mayer, C., Bachler, M., Wassertheurer, S., Pinho, A.J., Koslicki, D.: On entropy-based data mining. In: [12], pp. 209–226 (2014)
14. Holzinger, A. (ed.): Machine Learning for Health Informatics. LNAI, vol. 9605. Springer, Cham (2016)
15. Manca, V.: Infobiotics: Information in Biotic Systems. Springer, Heidelberg (2013)
16. Manca, V.: Grammars for discrete dynamics. In: [14], pp. 37–58. Springer, Heidelberg (2016)
17. Manca, V.: The principles of informational genomics. Theoret. Comput. Sci. (2017)
18. Manca, V.: An informational proof of H-theorem. Open Access Library (Modern Physics) 4, e3396 (2017)
19. Sharp, K., Matschinsky, F.: Translation of Ludwig Boltzmann's Paper "On the Relationship between the Second Fundamental Theorem of the Mechanical Theory of Heat and Probability Calculations Regarding the Conditions for Thermal Equilibrium". Entropy **17**, 1971–2009 (2015)
20. Schrödinger, E.: What is Life? The Physical Aspect of the Living Cell and Mind. Cambridge University Press, Cambridge (1944)
21. Shannon, C.E.: A mathematical theory of communication. Bell. Sys. Tech. J. **27**, 623–656 (1948)
22. Schervish, M.J.: Theory of Statistics. Springer, New York (1995)
23. Turing, A.M.: On computable numbers, with an application to the entscheidungsproblem. Proc. London Math. Soc. **42**(1), 230–265 (1936)
24. Wheeler, J.A.: Information, physics, quantum: The search for links. In: Zurek, W.H. (ed.) Complexity, Entropy, and the Physics of Information. Addison-Wesley, Redwood City (1990)
25. Wiener, N.: Cybernetics Or Control and Communication in the Animal and the Machine. Hermann, Paris (1948)

Beyond Volume: The Impact of Complex Healthcare Data on the Machine Learning Pipeline

Keith Feldman[1], Louis Faust[1], Xian Wu[1], Chao Huang[1], and Nitesh V. Chawla[1,2(✉)]

[1] University of Notre Dame, Notre Dame, IN, USA
{kfeldman,lfaust,xwu9,chuang7,nchawla}@nd.edu
[2] Indiana Biosciences Research Institute, Indianapolis, IN 46202, USA

Abstract. From medical charts to national census, healthcare has traditionally operated under a paper-based paradigm. However, the past decade has marked a long and arduous transformation bringing healthcare into the digital age. Ranging from electronic health records, to digitized imaging and laboratory reports, to public health datasets, today, healthcare now generates an incredible amount of digital information. Such a wealth of data presents an exciting opportunity for integrated machine learning solutions to address problems across multiple facets of healthcare practice and administration. Unfortunately, the ability to derive accurate and informative insights requires more than the ability to execute machine learning models. Rather, a deeper understanding of the data on which the models are run is imperative for their success. While a significant effort has been undertaken to develop models able to process the volume of data obtained during the analysis of millions of digitalized patient records, it is important to remember that volume represents only one aspect of the data. In fact, drawing on data from an increasingly diverse set of sources, healthcare data presents an incredibly complex set of attributes that must be accounted for throughout the machine learning pipeline. This chapter focuses on highlighting such challenges, and is broken down into three distinct components, each representing a phase of the pipeline. We begin with attributes of the data accounted for during preprocessing, then move to considerations during model building, and end with challenges to the interpretation of model output. For each component, we present a discussion around data as it relates to the healthcare domain and offer insight into the challenges each may impose on the efficiency of machine learning techniques.

Keywords: Healthcare informatics · Machine learning · Knowledge discovery

1 Introduction

Only in its infancy as a digital entity, the healthcare industry has undergone a significant transition over the past decade from a paper-based domain to one

© Springer International Publishing AG 2017
A. Holzinger et al. (Eds.): Integrative Machine Learning, LNAI 10344, pp. 150–169, 2017.
https://doi.org/10.1007/978-3-319-69775-8_9

operating primarily through a digital medium. Beyond the logistical benefits of maintaining and organizing patients' medical records, the ability to quickly identify and process information from millions of patient records, laboratory reports, imaging procedures, payment claims, and public health databases has brought the industry to the precipice of a significant change. Namely, the opportunity to utilize data science and machine learning methodologies to address problems across the practice and administration of healthcare.

In fact, utilization of such analytic techniques has provided a foundation on which models of personalized and predictive care have emerged [1]. These models represent a myriad of opportunities from improved patient stratification, to identifying novel disease comorbidities and drug interactions, to the prediction of clinical outcomes [2]. However, while such applications hold great promise for the healthcare industry, the application of machine learning methodologies faces a significant set of obstacles intrinsic to the data being evaluated and the population from which the data is drawn.

Since its entrance into the digital era, the increasing scale and scope of data has placed great emphasis on the advent of *Big Data* in healthcare and the challenges that come with it. With an estimated 150 exabytes of data generated by 2011, early work addressed the challenges of processing data at such a scale [3]. However, it is important to remember that Big Data is defined by more than just size, but rather by what are colloquially known as the four V's (The Volume, or quantity of data available. The Velocity, or speed at which the data is created. The Variety of the data elements available. And the Veracity, or inherent truthfulness of data itself) [4]. With advancements to the theoretical underpinning and practical implementations of machine learning algorithms providing the ability to consume and analyze even the largest clinical and biomedical datasets, the challenge now falls not to the size of the data, but its complexity.

In stark contrast to the idealistic data on which machine learning algorithms are theoried, healthcare data is inherently fragmented, noisy, high-dimensional, and heterogeneous. With the influx of data from an increasingly varied set of sources, it has become clear that effective utilization of these techniques will require more than accessibility of data or ability to execute Big Data analytics. As clinical research becomes increasingly intertwined with the statistical methodologies of data science, effective applications require an awareness to the mechanisms by which the data is created, processed, and analyzed.

To this end, the following chapter will address the complexities of healthcare data as they impact the machine learning algorithms which consume them. Broadly, we break such work into three major categories, each representing a component of the machine learning pipeline, as seen in Fig. 1. Beginning with preprocessing, we will discuss attributes of the data itself through the concepts of noise, missingness, and variability in language. We will then move to the modeling phase, discussing considerations such as the heterogeneity of data sources, sparsity and class imbalance. Finally, we will look to the model output, discussing the concepts of validation and verification. We will conclude with some general recommendations and a review of the open problems.

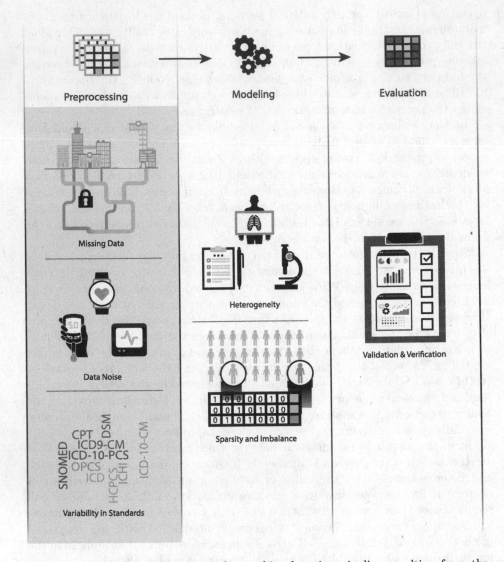

Fig. 1. Fundamental challenges to the machine learning pipeline resulting from the complexity of healthcare data

2   Glossary and Key Terms

Preprocessing: A process intended to address the noisy, missing, and inconsistent properties of real-world data, improving data quality prior to modeling. Preprocessing is often the first step in the machine learning pipeline and is characterized by techniques such as cleaning, integration, reduction, and transformation [5].

Modeling: The second stage of the machine learning pipeline, modeling focuses on the construction of statistical, probabilistic models intended to learn representations of the vast amounts of data collected. Such models are used to detect patterns in data and potentially use the patterns to predict future data [6].

Evaluation: Performed on the artifacts produced by modeling, evaluation forms the final stage of the pipeline, establishing the model's predictive efficacy, complexity, technical correctness, and ease with which it can be understood [7].

Validation: The process of evaluating a model in its ability to accurately represent the observed system [8].

Verification: The process of evaluating whether data manipulation and model construction were accomplished with technical correctness [8].

Medical Coding: The systematic classification of data into alphanumeric codes for the identification of diagnoses, procedures, medications, laboratory tests, and other clinical attributes [9].

Sparsity: Occurs when only a small percentage (typically <1%) of attributes for an instance are non-zero [10].

Concept Drift: The notion that inputs or outcomes related to a model may change overtime in unexpected manners reducing the accuracy of models as the data streams change [11].

3 Preprocessing

Just as clinicians require quick and accurate information to provide care at the highest level, the need to collect and produce high quality data has become paramount for applications of machine learning as they continue to integrate into aspects of care pertaining to health and human behavior. However, while the goal is clear, the rapid influx of new data, and the evolving nature of healthcare itself offers a significant set of challenges. In this regard, the following sections present an overview to the considerations of preprocessing data collected across the healthcare domain.

3.1 Manifestation in Healthcare

While the challenges to preprocessing are present in many domains, the dynamics of healthcare necessitate that care be taken to address a number of biological, computational, and representational aspects of data. These can range from the filtration of noise, to the need to navigate a multitude of coding standards. The following sections will begin by highlighting scenarios from which these challenges arise.

Noise. The presence of incorrect or irrelevant data, otherwise known as noise, represents a fundamental component of working with any real-world data. Healthcare is no exception, and arising from an imperfect data collection process, common occurrences of noise can include missing values, misspellings, abbreviations, misfielded values, word transpositions, and duplicated or conflicting records [12]. Though, the presence of noise stems from more than just data quality issues, it may also arise from the natural variation among individuals. Given a population of instances, a small sample may appear inconsistent with the rest i.e. "outliers".

Next, it is important to remember that noise is present not only in the recording of data, but in its measurement as well. Healthcare is currently entering uncharted territory. While traditionally the generation of health data was confined within the walls of a clinical setting, advancing technology has allowed for collection from a variety of sources. These range in complexity from personal health tools, to clinically focused devices, to total wireless sensor networks, to home monitoring systems [13–15]. However, with development from a number of manufacturers, utilizing a range of algorithmic techniques for their data collection and approximation, the quality of this data has been drawn into question [16,17]. A scenario highlighted by Bland and Altman, who note that "several measurements of the same quantity on the same subject will not in general be the same. This may be because of natural variation in the subject, variation in the measurement process, or both" [18].

Finally, we find that beyond the collection and recording of health data, there exists a more complex source of noise known as artifacts. Artifacts result not from data collection or variability in subjects, but from the physiological processes which generate the data itself, manifesting as what appears to be normal data. Although, in reality, such feature values are not generated by the intended source (e.g. electrical signals from the brain collected by an electroencephalogram). Instead, this data is generated by alternative biological mechanisms including cardiac, glossokinetic, muscle, eye movements, respiratory and pulse variations [19].

Missingness. The occurrence of missing data is an almost unavoidable problem for any domain, including healthcare [20]. Missing data can result from a number of processes, ranging from fundamental attributes of data collection to the inherent ambiguity and variability of an individual's health condition. At the most basic level, as with all studies that involve the collection of information from individuals, there exists the possibility of missing data attributes due to a subjects failure to respond completely, as well as the inability to assess all possible clinical and social attributes as they pertain to each individual. Additionally, missing data is not restricted to particular attributes, but can arise on a broader scale with attrition of an entire instance during longitudinal data collection. In addition to the to the lack of data, such a scenario presents difficultly during processing as the reason for dropping may be linked to attributes of the study design, a trend which may go unnoticed without closer investigation [21].

The evolving digital nature of healthcare presents its own set of challenges in regards to the presence of missing data. From a collection standpoint, monitoring devices may fail or become disconnected, data may become corrupt, or compatibility issues may result in the inability to collect data, resulting in large gaps of the recorded data [22].

Moreover, even in the scenario in which data collection occurs as expected, missingess can take other forms. Looking to the frequency of clinical encounters, it may occur from a temporal standpoint. With the exception of some critical care, patients are rarely under continuous observation, instead, many may meet with their physicians as infrequently as twice a year. These gaps in observation and records may allow for fluctuations in health to go undocumented, leaving only brief snapshots of the patients condition.

Further, missingness can occur due to the fragmented nature of the entities collecting the data. Healthcare data comes not only from hospitals and primary care centers, but a variety of sources, be that specialists visited, community programs, or even physical trackers [14,23]. However, despite the various data sources collecting data relevant to the overall profile of an individuals health, data integration and sharing considerations often provide only a small portion of data to any one source.

Finally, it is important to note a distinction between missing data and negative values as it relates to a perhaps non-traditional concept of missingness. Unlike domains such as retail, where the purchase of an item can be represented in a binary fashion (purchased or not), the lack of affirmation for a particular entity in healthcare data does not necessitate a negative case. Looking to disease diagnoses, a patient may in fact have a particular condition, for which they are never formally diagnosed, or for which the diagnosis code is not recorded, as is the case with often under-reported diagnoses such as obesity [24].

Variability in Language. Another challenge in processing healthcare data stems not from a function of its quality, but from its representation. In an effort to quantify and standardize the vast set of possible conditions, procedures, and clinical elements, a myriad of medical coding schemes have been developed, including the International Classification of Diseases (ICD), Systematized Nomenclature of Medicine (SNOMED), Current Procedural Terminology (CPT), Healthcare Common Procedure Coding System (HCPCS), LOINC, Europe's Classification of Surgical Operations and Procedures (OPCS), and the Diagnostic and Statistical Manual of Mental Disorders (DSM) to name a few. In fact, as the number of standards continues to increase there is a considerable amount of overlap between them. As a result, effective processing of such data must take care to consider the possibility where the same attribute may be represented in multiple ways. This situation is exacerbated by the nature of healthcare systems, where in response to documentation or reporting standards, multiple coding standards may be used even within the same institution.

Not only does variability arise from the use of different coding standards, but from emerging diversity as these standards are revised and updated. As an example, the ICD's latest revision (ICD-10) brought with it roughly 55,000 new diagnostic codes and over 68,000 new procedural codes [25]. Although this increased feature space allows for representation of conditions at a much greater specificity, coalescing codes across revisions during processing presents a significant challenge. Further, although mappings have been created to assist in the transition between codes, they are not universal and often incomplete. The Workgroup for Electronic Data Exchange suggest "healthcare organizations use these mappings as starting points to develop their own, more precise data crosswalk applications between ICD-9 and ICD-10 codes" [26].

While variability is clearly a product of the expansive set of coding standards and their revisions, it also results from the methodology of medical coding itself. Medical coding is a subjective process, the accuracy of which has been shown to be dependent on the clinical record of the condition observed, as well as the interpretation of the diagnostic codes themselves [27]. While it may be straightforward for simple cases where a patient is assigned a single diagnosis, inconsistencies from coders and institutions have been found to increase with the complexity of a patient's condition, specifically when they receive multiple diagnoses [28].

3.2 Implications to Machine Learning

Noted by Cortes et al., "insufficiencies of the data limit the performance of any learning machine or other statistical tool constructed from and applied to the data collection - no matter how complex the machine or how much data is used to train it" [29]. As a result, it is imperative to understand not only the processes from which preprocessing challenges arise within the healthcare domain, but also the implications to the preprocessing phase of the machine learning pipeline. A discussion to each consideration can be found in the sections below.

Noise. As applications of machine learning continue to expand into new aspects of healthcare, the processing of noisy data has become a central component of many works. From a theoretical perspective, prior work has established fundamentals of what defines learning, and concepts of model consistency. Together, these constructs help illustrate how failure to process noise in data can cause difficulties in constructing a model that accurately reflects the population from which the data is drawn, negatively impacting generalizable performance [30–32].

While much of the standard noise can be attributed to data quality issues, it is important to highlight the need for data understanding in the preprocessing step. In particular, with relation to outliers. There are over 100 different discordancy/outlier tests whose use can depend on factors such as data distribution, whether distribution parameters are known, and even the number and type of the expected outliers [33]. As such, preprocessing noisy data presents a significant challenge, as incorrectly applying data cleaning techniques can result in large variations in the finalized dataset.

Looking to other sources of noise, the nature of potential physiological artifacts requires additional considerations during data preprocessing. The presence of an artifact does not necessitate the value be incorrect, though similarly to outliers, failure to remove such data has far reaching implications as such data presents "a milder form of training data error that can cause reduced accuracy" [34]. However, unlike outliers, these artifacts are often difficult to broadly and statistically discern from true signals without clinical insight.

Finally, in addition to the data quality concerns already discussed, noise resulting from the variability of the systems which collect data presents a distinct concern during processing. Due to the resulting intra-instance variability, there exists the case in which two instances with identical feature values present two different classes or outcomes. Whereas such instances are often removed, within healthcare, such a scenario is quite common and may represent a legitimate aspect of variability with a patient's health. This in turn, introduces a considerable amount of uncertainly into the system, representing an inherent problem to separability.

Missingness. Just as the numerous sources of noise present a challenge to the effective processing of healthcare data, so too do the many forms in which missing data can manifest across the domain. At an attribute-level, data is typically classified as missing in one of three forms: completely at random (MCAR), at random (MAR), and not at random (MNAR). Although all forms of missingess present a concern, the various forms of missingness can present significantly different considerations during the processing of a dataset. While data missing completely at random (MCAR) presents minimal concern to the underlying distribution, allowing for data to be dropped or imputed without worry of introducing additional biases, such a scenario is often unrealistic. Rather, data is typically missing due to an underlying, sometimes unobserved, pattern known as missing at random (MAR) or missing not at random (MNAR), each of which may require techniques such as maximum likelihood estimation or multiple imputation to help address the inherent bias they present to the data collected [35–37]. In comparison, both methods tend to yield similar results when implemented in the same way, however, performance gains in efficiency and reduced bias regarding these methods relate to the inclusion of auxiliary variables [38]. The inclusion of data to these methods, even that which is irrelevant to the objective at hand, suggests the amount of data included in these methods is of equal importance to the methods themselves.

Beyond the type of missingness, the quantity of missing information further influences the preproccessing of data. With respect to the occurrences of large temporal gaps, we find that although mathematically we may be able to impute, model and predict estimations of missing values during processing there is no guarantee the values computed accurately reflect the true condition of the individual during that time period. This consideration is particularly relevant in light of the common scenario where data is collected during a subjects clinical encounters, each of which may occur months apart.

Finally, building on the concerns of temporal missingess, commonly associated with longitudinal studies, missingness by attrition, presents a number of additional considerations to effective preprocessing of data. Work by Graham suggests attrition-related missingness focuses on the program (or treatment) P, the dependent Y, and the interaction between these two: PY [39]. Just as the MCAR/MAR/MNAR nomenclature provides a roadmap to the appropriate preprocessing techniques, identifying and assessing which of the possible combinations of these three factors causes missingess to arise presents a critical step in improving the ability to address bias during the processing of such data.

Variability in Language. The variability brought on by the breath of coding standards presents a fundamental obstacle in the effective preprocessing of healthcare data. Although an underlying condition may be the same across two distinct representations, with the multitude of values across each of the different coding standards, it has become nearly impossible to accurately create a comprehensive mapping to translate between each standard. However, such a mapping is critical for unifying disparate data sources during the processing stage of the machine learning pipeline.

Further, although data may stay consistent with respect to a single coding standard, temporal changes in how these codes are assigned can still occur as a result of changing regulations, or even revisions within the standard [40]. Presenting similar obstacles as with multiple coding standards, these changes, more formally defined by the notion of concept drift, cause models built on old data to become inconsistent with new data as the models inputs and target variable change over time [11].

While such a shift is extremely difficult to identify, it is critically important, as such discrepancies make it not only difficult to understand values, but have the potential to add ambiguity during its processing [41]. In particular, changes in the code frequencies, which are often used during preprocessing and data exploration, may be attributed to other clinical attributes, rather than the true shift in language. For example, an individual with a chronic illness may have a record with multiple representations due to changes in how the illness was labeled over time. When such changes go unaccounted for, the record may be perceived as having three distinct illnesses instead of one.

4   Modeling

To this point we have discussed intrinsic characteristics of data, those properties which influence the statistical foundations guiding machine learning theory. However, we now look further, not to the properties of the data, but to the mechanisms through which the data is consumed and represented to build effective machine learning models. Such attributes range from high-level aspects of integrating heterogeneous data types, to low-level considerations when representing an increasingly expansive feature space.

4.1   Manifestation in Healthcare

As before, we will begin with an outline of the processes within the healthcare domain from which such considerations arise. An overview of each can be found in the respective sections to follow.

Heterogeneity. Drawn from multiple sources and encompassing multiple modalities, healthcare data represents a remarkably heterogeneous set of data types and sources. Perhaps the most prominent examples can be found within the wealth of clinical data now digitalized as a result of EMR integration across healthcare practices. Typically, such data is broken into structured data including diagnosis codes, procedural information, medication data, laboratory test results, data recorded directly from patient's bedside monitors, and unstructured data such as images and clinical text [42]. However, data can also include patient demographics, financial claims, and more recently, genomic sequencing and other omics data, each of which may require different considerations as they are processed during modeling.

Although electronic health records are perhaps the most well-known source of data, health-related data can be collected, inferred, and analyzed from a number of indirect sources. These can include common population health and reporting fields such as the census bureau and the department of labor statistics, as well as less obvious sources such as the location of fresh food sources in a city. It is also worth noting the number of external data is only expected to increase, as shifts in the regulatory landscape of the healthcare industry have advanced the collection and analysis of population health data though a number of initiatives [43].

Finally, it is important to note that heterogeneity can exist even within data of the same type. As an example, through prior work our group has established fundamental differences between clinical notes based on the clinical occupation of those who write them [44]. While important for the processing of clinical text, the establishment of such heterogeneity impresses a deeper need for an awareness of not only the types of data we process, but the varied sources of data from which models are constructed.

Dimensionality, Sparsity and Imbalance. Beyond the variety of sources generating data, the digitalization of healthcare data has resulted in a significant increase in the number of features able to be extracted with respect to an instance, i.e. its dimensionality. Although the high-dimensional data resulting from the processing of unstructured images and text has become commonplace, advancements in clinical and computational technology now allow for improved analysis of biological processes at their most basic level. As an example, resulting from the increasing affordability of genomic sequencing, we have witnessed a rise of genome-wide association studies, which aim to represent and identify associations between the over 10 million common single-nucleotide polymorphisms (SNPs) in the human genome [45].

However, such expansive feature sets are not only a result of biological mechanisms, but of artificial constructs used to structure the data itself. In particular, we look to the coding standards used to represent attributes of a patient's condition and care. As noted prior, there exist a multitude of standards, each potentially representing tens of thousands of unique codes.

In fact, it is the expansiveness of the resulting feature space that leads us to the next aspect of healthcare data that has been shown to impact modeling: sparsity. Although we may be able to capture, code, and quantify an increasingly large feature set, only a small subset of features are often recorded or relevant for a particular individual. Such a point can be best illustrated with the understanding that it is highly improbable a patient will record more than a fraction of the over 100,000 diagnoses and procedures that can now be discretely represented through the ICD-10 standard [25].

Finally, taking the considerations of dimensionality and sparsity to the next logical step, we find the concept of imbalance. Building on the notion that for any single instance, the data captured likely represents only a sparse set of values with respect to the possible set of data elements, we must acknowledge that these same features are often used as the response variable for many machine learning applications. Whether the prediction of a future diagnosis, or the readmission probability of patient, effective utilization of data in which only a minority of individuals present the attribute of interest often requires additional processing, or specialized models.

4.2 Implications to Machine Learning

Having illustrated a number of processes from which data challenges arise within the healthcare domain, we again address how these challenges, at the modeling stage, impact machine learning algorithms.

Heterogeneity. From a technical perspective, one of the primary considerations in the application of machine learning methodologies to the increasingly heterogeneous healthcare data space comes with the acknowledgment that the data captured across each source may span a range of data types. Such data can represent categorical/discrete (laboratory SRI values), ordinal (pain scales), or continuous (medication dosages) values. However, as noted by Lewis et al., "traditional statistical methods that assume Gaussian distributions, or engineering methods that assume vector or matrix input do not obviously generalize to datasets comprised of variable-length strings, vectors of real numbers, trees and networks" [46].

Further, the heterogeneity of the data sources themselves present a challenge for learning algorithms. As the extent of available healthcare data continues to increase, we must also consider the implications of integrating disparate sources. There, of course, exist the practical concerns including record matching and differences in terminology and standards between systems, where a patient may include their middle name on one form but not another or systems may record height as meters or feet. However, the modeling of heterogeneous integrated data sources presents a number of unique concerns, including the ability to reconcile "dirty" data such as incompatible test results, changes in coded data, and the need to ensure trust between systems that share sensitive data [47].

Finally, at its core, the siloed data sources present a deeper systemic issue. The lack of a unique identifier to track an individual throughout the various components of the healthcare system often present an incomplete view of any individual's health data [48]. While the missing data itself can present concern, the impact of this fragmentation compounds during the integration of multiple data sources. Incorrectly associating records of one patient from two hospitals, records between a primary and specialist, or multiple instances of the same record has serious implications to the machine learning algorithms used to analyze the data. At a basic level, this provides duplicate data that can bias the underlying distributions. While at a higher-level, this removes a true independence assumption, potentially biasing performance measures by splitting what appears to be unique instances amongst the train and test sets during evaluation.

Dimensionality, Sparsity and Imbalance. With the considerable advancement of computing systems and hardware over the past few decades, the ability to store, represent, and manipulate high-dimensional data has become commonplace. However, the appropriate utilization of such data in the machine learning pipeline warrants additional consideration. In particular, the scenario often known as the *curse of dimensionality*, in which the number of features approaches or exceeds the number of instances, presents a considerable obstacle to fundamental machine learning theory. One of the most direct impacts results from the emergence of spurious correlation, where many uncorrelated random variables may have high sample correlations [49]. While more indirectly such a scenario has been shown to breakdown asymptotic theory, preventing the unique estimation of parameters due to occurrence of singular matrices [50].

Further, the sparsity of feature values presents an additional concern to the development of generalizable models. The identification of latent interactions between features is often not complete, as combinations between all features is rarely captured. While in theory, such an issue can be alleviated with the collection of a large dataset, as feature spaces become ever-larger, the collection of sufficient data is not often feasible due to logistical and economical considerations. While many newer machine learning techniques have looked to sparsity as a foundation for techniques that address the increasing dimensionality using various greedy algorithms, little theoretical support is currently available for such techniques [51].

Finally, as noted prior, the presence of sparsity often leads to the traditional class imbalance problem, where the attribute of interest is possessed by only a subset of instances. The implications of such imbalance on statistical learning tasks have been well established in a number of prior works [52]. However it is worth noting a few examples of how the presence of imbalance can impact the modeling of data. From a logistical point of view, a few noisy instances can degrade the identification of the minority class, restricting it to fewer examples to train with. Whereas from a more theoretical presumptive the use of global performance measures that guide the learning process may optimize parameters and decision boundaries in favor of the majority class. Classification rules that predict the positive class are often highly specialized resulting in low coverage of instances across the dataset and may be discarded in favor of providing more general rules [53].

5   Evaluation

"All models are wrong, some are useful" [54]. The provocative and now-famous quote by George Box eloquently provides a fundamental premise of machine learning. The understanding that models may not truly capture the complexity of a system, but rather provide an effective approximation of its observable attributes. This concept has since been defined more formally by Oreskes et al., stating "Model results may or may not be valid, depending on the quality and quantity of the input parameters and the accuracy of the auxiliary hypotheses" [55]. In actuality, what these sentiment capture is the need to construct a model in such a way that it accurately represents the system (validation) and accounts for the technical correctness of the model itself (verification) [8].

5.1   Manifestation in Healthcare

Together, validation and verification represent a critical aspect of computational tools such as machine learnings impact on our society. However, accurate assessment of either measure proves a nontrivial task in its own right. Difficulties associated with validation and verification are grounded by two district ideas which govern both measures. First, that there exists a ground truth, and second that any ground truth provided is correct. While both of these assumptions are difficult to guarantee for any real-world data, they are particularly relevant to the variability of data across the healthcare domain.

The first concern represents a variant of the partially labeled data problem, formally defined by Szummer [56]. As noted in the *Preprocessing* section, although missing values are commonly treated as negative, such an assumption is dangerous. Diagnoses that are never recorded or identified can present the scenario in which two individuals may appear the same from a record standpoint, but in fact present significantly different clinical outcomes. Further, there is a well-established issue in the ability to record clinical conditions completely within any particular coding languages [57]. Thus, in an effort to address this lack of a ground truth for missing diagnoses, many machine learning approaches utilize only positive entities, in essence, formulating a one-class problem.

This in turn leads to the second point of concern, where even in the scenario in which a diagnosis is recorded, such data is not guaranteed to be correct. As diagnosis information is often obtained though a diagnostic test, it is important to remember these tests are subjective. Each test is associated with its own performance range quantified by metrics such as sensitivity and specificity, predictive values, chance-corrected measures of agreement, or likelihood ratios [58]. Without the ability to identify false positives/negatives, or to link to follow-up tests or corrections, these results represent a significant source of error which can propagate through an algorithmic model.

5.2 Implications to Machine Learning

The challenges associated with the validation and verification of healthcare data impact far more than application-specific performance, reaching to the intermittent steps of the algorithms underlying the solutions themselves. In particular, the complexities of healthcare data impacts both the internal distance metric utilized, as well as the generalized optimization problem of parameter tuning.

Although simplistic in definition, the notion of distance represents a fundamental attribute in the execution of machine learning algorithms. The ability to quantify a pair of similar or dissimilar points is a concept utilized in determining decision boundaries, as well as updates for weights. Such a concept has become increasingly important as a result of the heterogeneity exhibited by the growing set of healthcare data discussed prior. While work by Brian Kulis highlights extensions of the metric learning problems to a variety of problems in computer vision, text analysis, and multimedia, these extensions represent nontrivial alterations to the concept of what constitutes *distance* [59].

From an optimization standpoint, in an effort to more accurately model complex real-world systems, an increasingly complex set of learning models have become available. Although these models have the potential to improve performance, there often exist a number of parameters (regularization strengths, number of weak learners, slack variables) that must be tuned to fit the data being modeled [60, 61]. While specifics of their implementation can vary greatly, to prevent the need to sweep the entirety of the parameter space, a model's parameters can be estimated through optimization techniques applied to an objective function of the users choosing. As there exist a range of possible optimizations, many stemming from a methodology known as gradient decent, an understanding of the data itself is paramount. Variability in how the specified objective function accounts for factors such as class imbalance or inter-feature correlation can result in markedly different results drawn from the same data [62–64].

Finally, it is important to note that validation and verification have often constituted major components in the assessment of model output, for both its generalizability and overall correctness. While critical to their real-world utility, these concerns directly impact the inferences drawn from model output, and as such, exist outside the scope of this chapter. An excellent survey addressing such items can be found in the work by Sokolova and Lapalme [65].

6   Open Problems

This work has served as a foundation highlighting a broad set of considerations that must be addressed as machine learning works to establish its place in the healthcare industry. However, as clinical practice, administration, and research becomes increasingly intertwined with the statistical methodologies of machine learning, there remains a number of open problems. In the sections to follow, we discuss a subset of the most pressing and active areas of research.

6.1 Temporal Relations

As healthcare data has undergone the transition from a paper-based entity to digital records, much of the focus has fallen to the purely technical aspects of storing, processing, and modeling such a complex set of variables. Amongst these considerations, however, we often forget to reflect on whether the data we consume accurately reflects the processes it captures. In particular, we find that an overwhelming majority of works have recorded and modeled the condition of an individual as a set of discrete observations.

Although such an approach allows for data to be easily consumed by traditional machine learning approaches, formalizing the presence of each entity recorded as a feature, such a representation is incomplete. It is clear observations such as a diagnosis or procedure must occur at a single point in time, however, it would be naive to believe that such elements of health occur in isolation. Rather, there exist temporal relations connecting them, representing the variable nature of an individual's health. These relations cannot be described by one feature or a single value, but require longitudinal observations with a series of values over time [66].

However, this is a nontrivial task from both a computational and clinical standpoint. Taking the example of a patients diagnosis history, the computational complexity of tracking the progression of multiple concurrent diagnoses can quickly become intractable on even the largest of systems, while from a clinical standpoint, many diagnoses have no direct progression, where others may split amongst a broad set of co-morbid diagnoses.

6.2 Alternative Representations

Building on the notion that continued improvement of machine learning approaches to problems in the healthcare domain may require a shift in our data organization, such as the ability to capture temporal relations, there exists a significant effort to employ varying frameworks to represent the complexity of healthcare data. Perhaps the most well explored representation can be found in the application of computational networks. Networks represent an established field of interdisciplinary research and present an effective method to capture the direct relation between two arbitrary features, be that a connection between individuals themselves, links between comorbid diagnoses, or genomic-phenotypic relations [67, 68]. Such a representation offers many attractive properties, such as the ability to alleviate sparsity by connecting only those elements associated with an instance, and the ability to represent heterogeneous data. However, the analytic methods applied to networks focus primarily on describing connectivity, with measures such as centrality, degree, and betweenness, rather than the generative or discriminative models constructed to describe the relations between various healthcare features.

In an effort to capture such relations, tensor representations have emerged. A tensor is a multidimensional array spanning an arbitrary number of dimensions, each representing a single feature or modality. Such a representation is advantageous to many areas of healthcare including the ability to capture a series of observations over time or integrate data across multiple experimental conditions and analyze them simultaneously [69]. Although tensors are only emerging in their application to healthcare, mathematical operations known as decompositions have demonstrated their value in discovering latent groups in each modality and identify group-wise interactions [70].

6.3 Integration with Clinicians and Clinical Workflows

Despite the immense technological advancements in the collection and processing of health data, it is important to remember that no system can succeed on its own. We would be remiss in failing to highlight that the impact of machine learning in healthcare cannot be discussed in isolation. Medical research is itself an evolving field, and an understanding of the biological processes being modeled requires a more technical approach. Truly capturing such phenomena will require close interdisciplinary collaborations with those individuals whose expertise lies in the exploration and discernment of healthcare.

It is perhaps more appropriate to view the continued development of healthcare informatics as part of the complex system encompassing clinical workflows. In relation to the collection and aggregation of data, it is important to note that any increase in data collection poses tangle logistic concerns to those individuals involved in the care of an individual. For example, while it may be more accurate to assess an individual's condition every minute, such granularity is often not feasible. As a result, there must be an increasing focus on developing innovative ways to utilize data collected as part of existing workflows, rather than demonstrating value with models requiring additional data elements. On the other end of the spectrum, in relation to model output, it is important to remember that regardless of the analytic approach used; a patient's treatment ultimately remains in the hands of their clinician. As such, there must be a concerted effort to provide appropriate context to the results. Designing approaches with the capability to quantify factors such as confidence, highlighting a systems strengths, and its weaknesses, for the individual consuming the information.

7 Conclusion and Future Outlook

Complexity comes in many forms, and beyond its sheer volume, healthcare data presents heterogeneous, high-dimensional, probabilistic, incomplete, uncertain, and noisy attributes. However, in conjunction with machine learning methodologies, the increasing availability of data has the potential to provide novel and actionable insights to the field of healthcare.

With this in mind, it is important to remember that although we may have reached a point computationally in which the manipulation of Big Data and execution of complex modeling techniques are possible, purely possessing such capability is not sufficient to ensure the realization of informatics true potential in healthcare. The ability to address such complexities and ensure the effective consumption of healthcare data into the machine learning pipeline relies on more than analytic capability: it relies on a deep understanding of the biological and clinical mechanisms through which the data has been generated.

From preprocessing, to modeling, to the interpretation of model output, this work has presented a general discussion regarding the fundamental challenges presented by such data with respect to the informatics pipeline. However, awareness of such an impact is only the first step. It is our hope that others will draw on these caveats and look to considerations in the design and implementation of new works, which blend together both technological capability and medical understanding to better serve those individuals in need.

References

1. Yoo, I., Alafaireet, P., Marinov, M., Pena-Hernandez, K., Gopidi, R., Chang, J.F., Hua, L.: Data mining in healthcare and biomedicine: a survey of the literature. J. Med. Syst. **36**(4), 2431–2448 (2012)
2. Jensen, P.B., Jensen, L.J., Brunak, S.: Mining electronic health records: towards better research applications and clinical care. Nat. Rev. Genet. **13**(6), 395–405 (2012)
3. Hughes, G.: How big is big data in healthcare. From a Shot in the Arm Blog (2011)
4. Raghupathi, W., Raghupathi, V.: Big data analytics in healthcare: promise and potential. Health Inf. Sci. Syst. **2**(1), 3 (2014)
5. Han, J., Pei, J., Kamber, M.: Data Mining: Concepts and Techniques. Elsevier (2011)
6. Murphy, K.P.: Machine Learning: A Probabilistic Perspective. MIT Press, Cambridge (2012)
7. Sammut, C., Webb, G.I.: Encyclopedia of Machine Learning. Springer Science & Business Media, New York (2011)
8. Kantardzic, M.: Data Mining: Concepts, Models, Methods, and Algorithms. Wiley, Chichester (2011)
9. Diamond, M.: Mastering Medical Coding. Elsevier Health Sciences (2013)
10. Tan, P.N., et al.: Introduction to Data Mining. Pearson Education India (2006)
11. Tsymbal, A.: The problem of concept drift: definitions and related work. Computer Science Department, Trinity College Dublin 106(2) (2004)
12. Rahm, E., Do, H.H.: Data cleaning: problems and current approaches. IEEE Data Eng. Bull. **23**(4), 3–13 (2000)
13. King, L.A., Fisher, J., Jacquin, L., Zeltwanger, P.: The digital hospital: opportunities and challenges. J. Healthc. Inf. Manag. JHIM **17**(1), 37–45 (2002)
14. Andreu-Perez, J., Leff, D.R., Ip, H.M., Yang, G.Z.: From wearable sensors to smart implants–toward pervasive and personalized healthcare. IEEE Trans. Biomed. Eng. **62**(12), 2750–2762 (2015)

15. Kidd, C.D., Orr, R., Abowd, G.D., Atkeson, C.G., Essa, I.A., MacIntyre, B., Mynatt, E., Starner, T.E., Newstetter, W.: The aware home: a living laboratory for ubiquitous computing research. In: Streitz, N.A., Siegel, J., Hartkopf, V., Konomi, S. (eds.) CoBuild 1999. LNCS, vol. 1670, pp. 191–198. Springer, Heidelberg (1999). doi:10.1007/10705432_17

16. Caceres, C.A.: Medical Devices-measurement, Quality Assurance, and Standards. Number 800. ASTM International (1983)

17. Koumoundouros, E.: Clinical engineering and uncertainty in clinical measurements. Australas. Phys. Eng. Sci. Med. 37(3), 467 (2014)

18. Bland, J.M., Altman, D.G.: Statistics notes: measurement error. BMJ 313(7059), 744 (1996)

19. Sethi, N., Sethi, J., Torgovnick, E., Arsura, E.: Physiological and non-physiological EEG artifacts. Internet J. Neuromonitoring 5(1) (2007)

20. Wood, A.M., White, I.R., Thompson, S.G.: Are missing outcome data adequately handled? A review of published randomized controlled trials in major medical journals. Clin. Trials 1(4), 368–376 (2004)

21. Little, R.J., D'agostino, R., Cohen, M.L., Dickersin, K., Emerson, S.S., Farrar, J.T., Frangakis, C., Hogan, J.W., Molenberghs, G., Murphy, S.A., et al.: The prevention and treatment of missing data in clinical trials. N. Engl. J. Med. 367(14), 1355–1360 (2012)

22. Marlin, B.M., Kale, D.C., Khemani, R.G., Wetzel, R.C.: Unsupervised pattern discovery in electronic health care data using probabilistic clustering models. In: Proceedings of the 2nd ACM SIGHIT International Health Informatics Symposium, pp. 389–398. ACM (2012)

23. Azarm-Daigle, M., Kuziemsky, C., Peyton, L.: A review of cross organizational healthcare data sharing. Procedia Comput. Sci. 63, 425–432 (2015)

24. Quan, H., Li, B., Duncan Saunders, L., Parsons, G.A., Nilsson, C.I., Alibhai, A., Ghali, W.A.: Assessing validity of ICD-9-CM and ICD-10 administrative data in recording clinical conditions in a unique dually coded database. Health Serv. Res. 43(4), 1424–1441 (2008)

25. International classification of diseases, (ICD-10-CM/PCS) transition, October 2015

26. Meyer, H.: Coding complexity: US health care gets ready for the coming of ICD-10. Health Aff. 30(5), 968–974 (2011)

27. Fisher, E.S., Whaley, F.S., Krushat, W.M., Malenka, D.J., Fleming, C., Baron, J.A., Hsia, D.C.: The accuracy of medicare's hospital claims data: progress has been made, but problems remain. Am. J. Public Health 82(2), 243–248 (1992)

28. MacIntyre, C.R., Ackland, M.J., Chandraraj, E.J., Pilla, J.E.: Accuracy of ICD-9-CM codes in hospital morbidity data, victoria: implications for public health research. Aust. N. Z. J. Public Health 21(5), 477–482 (1997)

29. Cortes, C., Jackel, L.D., Chiang, W.P., et al.: Limits on learning machine accuracy imposed by data quality. KDD 95, 57–62 (1995)

30. Vapnik, V.N., Vapnik, V.: Statistical Learning Theory, vol. 1. Wiley, New York (1998)

31. Kearns, M.J., Vazirani, U.V.: An Introduction to Computational Learning Theory. MIT press (1994)

32. Sessions, V., Valtorta, M.: The effects of data quality on machine learning algorithms. ICIQ 6, 485–498 (2006)

33. Knorr, E.M., Ng, R.T., Tucakov, V.: Distance-based outliers: algorithms and applications. VLDB J. Int. J. Very Large Data Bases 8(3–4), 237–253 (2000)

34. Bacioiu, A.S., Sauntry, D.M., Boyle, J.S., Wong, L.C.W., Leonard, P.F., Chandrasekar, R.: Method and apparatus for analysis and decomposition of classifier data anomalies. US Patent 7,426,497, 16 September 2008
35. Little, R., Rubin, D.: Statistical analysis with missing data (1987)
36. Arbuckle, J.L., Marcoulides, G.A., Schumacker, R.E.: Full information estimation in the presence of incomplete data. In: Advanced Structural Equation Modeling: Issues and Techniques, vol. 243, p. 277 (1996)
37. Rubin, D.B.: Multiple Imputation for Nonresponse in Surveys, vol. 81. Wiley (2004)
38. Collins, L.M., Schafer, J.L., Kam, C.M.: A comparison of inclusive and restrictive strategies in modern missing data procedures. Psychol. Methods **6**(4), 330 (2001)
39. Graham, J.W.: Missing data theory. In: Graham, J.W. (ed.) Missing Data, pp. 3–46. Springer, New York (2012). doi:10.1007/978-1-4614-4018-5_1
40. Rector, A.L., Brandt, S.: Why do it the hard way? The case for an expressive description logic for snomed. J. Am. Med. Inform. Assoc. **15**(6), 744–751 (2008)
41. Lindenauer, P.K., Lagu, T., Shieh, M.S., Pekow, P.S., Rothberg, M.B.: Association of diagnostic coding with trends in hospitalizations and mortality of patients with pneumonia, 2003–2009. JAMA **307**(13), 1405–1413 (2012)
42. Weber, G.M., Mandl, K.D., Kohane, I.S.: Finding the missing link for big biomedical data. JAMA **311**(24), 2479–2480 (2014)
43. Stoto, M.A.: Population health in the Affordable Care Act Era, vol. 1. Academy-Health, Washington, DC (2013)
44. Feldman, K., Hazekamp, N., Chawla, N.V.: Mining the clinical narrative: all text are not equal. In: 2016 IEEE International Conference on Healthcare Informatics (ICHI), pp. 271–280. IEEE (2016)
45. Visscher, P.M., Brown, M.A., McCarthy, M.I., Yang, J.: Five years of GWAS discovery. Am. J. Hum. Genet. **90**(1), 7–24 (2012)
46. Lewis, D.P., Jebara, T., Noble, W.S.: Support vector machine learning from heterogeneous data: an empirical analysis using protein sequence and structure. Bioinformatics **22**(22), 2753–2760 (2006)
47. Diamond, C.C., Mostashari, F., Shirky, C.: Collecting and sharing data for population health: a new paradigm. Health Aff. **28**(2), 454–466 (2009)
48. Hillestad, R.: Identity crisis: an examination of the costs and benefits of a unique patient identifier for the US health care system. Rand Corporation (2008)
49. Fan, J., Han, F., Liu, H.: Challenges of big data analysis. Natl. Sci. Rev. **1**(2), 293–314 (2014)
50. Johnstone, I.M., Titterington, D.M.: Statistical challenges of high-dimensional data (2009)
51. Lafferty, J.D., Wasserman, L.: Challenges in statistical machine learning. Statistica Sinica **16**, 307 (2006)
52. He, H., Garcia, E.A.: Learning from imbalanced data. IEEE Trans. Knowl. Data Eng. **21**(9), 1263–1284 (2009)
53. López, V., Fernández, A., García, S., Palade, V., Herrera, F.: An insight into classification with imbalanced data: empirical results and current trends on using data intrinsic characteristics. Inf. Sci. **250**, 113–141 (2013)
54. Box, G.E.: Robustness in the strategy of scientific model building. Robust. Stat. **1**, 201–236 (1979)
55. Oreskes, N., Shrader-Frechette, K., Belitz, K., et al.: Verification, validation, and confirmation of numerical models in the earth sciences. Science **263**(5147), 641–646 (1994)
56. Szummer, M.O.: Learning from partially labeled data. PhD thesis, Massachusetts Institute of Technology (2002)

57. Gensinger Jr., R.A.: Analytics in Healthcare: An Introduction. HIMSS (2014). CPHIMS, FHIMSS
58. Glas, A.S., Lijmer, J.G., Prins, M.H., Bonsel, G.J., Bossuyt, P.M.: The diagnostic odds ratio: a single indicator of test performance. J. Clin. Epidemiol. 56(11), 1129–1135 (2003)
59. Kulis, B., et al.: Metric learning: a survey. Found. Trends® Mach. Learn. 5(4), 287–364 (2013)
60. Arcuri, A., Fraser, G.: Parameter tuning or default values? An empirical investigation in search-based software engineering. Empir. Softw. Eng. 18(3), 594–623 (2013)
61. Hoos, H.H.: Automated algorithm configuration and parameter tuning. In: Hamadi, Y., Monfroy, E., Saubion, F. (eds.) Autonomous Search, pp. 37–71. Springer, Heidelberg (2011). doi:10.1007/978-3-642-21434-9_3
62. Kelley, C.T.: Iterative methods for optimization. SIAM (1999)
63. Sra, S., Nowozin, S., Wright, S.J.: Optimization for Machine Learning. MIT Press (2012)
64. Lange, K., Chi, E.C., Zhou, H.: A brief survey of modern optimization for statisticians. Int. Stat. Rev. 82(1), 46–70 (2014)
65. Sokolova, M., Lapalme, G.: A systematic analysis of performance measures for classification tasks. Inf. Process. Manage. 45(4), 427–437 (2009)
66. Zhao, J., Papapetrou, P., Asker, L., Boström, H.: Learning from heterogeneous temporal data in electronic health records. J. Biomed. Inform. 65, 105–119 (2017)
67. Carter, H., Hofree, M., Ideker, T.: Genotype to phenotype via network analysis. Curr. Opin. Genet. Dev. 23(6), 611–621 (2013)
68. Feldman, K., Stiglic, G., Dasgupta, D., Kricheff, M., Obradovic, Z., Chawla, N.V.: Insights into population health management through disease diagnoses networks. Sci. Rep. 6, Article no. 30465 (2016)
69. Hunyadi, B., Van Huffel, S., De Vos, M.: The power of tensor decompositions in biomedical applications (2016)
70. Luo, Y., Wang, F., Szolovits, P.: Tensor factorization toward precision medicine. Brief. Bioinform. 18(3), 511–514 (2017)

A Fast Semi-Automatic Segmentation Tool for Processing Brain Tumor Images

Andrew X. Chen[1] and Raúl Rabadán[1,2(✉)]

[1] Department of Systems Biology, Columbia University Medical Center,
1130 St. Nicholas Avenue, New York, NY 10032, USA
{ac3957,rr2579}@cumc.columbia.edu
[2] Department of Biomedical Informatics, Columbia University Medical Center,
1130 St. Nicholas Avenue, New York, NY 10032, USA

Abstract. Segmentation, the process of delineating boundaries and features within images, is a vital part of both the clinical assessment and the computational analysis of brain cancers. Here, we provide an open-source algorithm (MITKats), built on the Medical Imaging Interaction Toolkit, to provide user-friendly and expedient tools for semi-automatic segmentation. To evaluate its performance against competing algorithms, we applied MITKats to MRIs of 38 high-grade glioma cases from publicly available benchmarks. The similarity of the segmentations to expert-delineated ground truths approached the discrepancies among different manual raters, the theoretically maximal precision. The average time spent on each segmentation was 5 min, making MITKats between 4 and 11 times faster than competing semi-automatic algorithms, while retaining similar accuracy. We conclude with remarks on the utility of segmentation for medical data analysis as well as its further challenges.

Keywords: Image processing · Semi-automatic segmentation · Brain tumor · Glioblastoma · MRI

1 Introduction and Motivation

Glioblastoma (GBM) is the most common primary brain malignancy, and the one with the most dismal prognosis [1]. Imaging, particularly magnetic resonance imaging (MRI), is the standard for diagnosing and assessing the disease [2]. The separation of tumor apart from healthy tissue within images (segmentation) is important for guiding therapy [3,4], determining prognosis [5,6], and assessing response [7,8]. Segmentation also serves as the first step in quantitatively analyzing GBM images, including machine learning [9–11], regression modeling [12,13], and clustering approaches [14]. These efforts have demonstrated utility in predicting tumor growth [15], molecular subtype and survival [16] of GBM patients. Given the recent availability of open cancer datasets [17] such as The Cancer Imaging Archive [18], there is a clear need for the development of efficient and accurate segmentation utilities, which will allow for the systematic quantification of images in association with clinical and molecular characteristics.

© Springer International Publishing AG 2017
A. Holzinger et al. (Eds.): Integrative Machine Learning, LNAI 10344, pp. 170–181, 2017.
https://doi.org/10.1007/978-3-319-69775-8_10

2   Glossary and Key Terms

GBM: glioblastoma, a brain tumor that is the most aggressive form of glioma. It is characterized by diffuse infiltration, regions of *necrosis* (dead tissue), and uniformly poor prognosis.

MRI: magnetic resonance imaging, a non-invasive medical technique often used to volumetrically visualize the brain as a series of image slices. Depending on the settings of the machine, images can be acquired in a variety of modalities such as T1, T1 with contrast (T1c), T2, and FLAIR. Gadolinium contrast is added in T1c to *enhance* (light up) disruptions of the blood-brain barrier.

Semi-automatic: An algorithm that combines user input with computer assistance, in contrast to fully manual or fully automated protocols.

Segmentation: The demarcation of images into distinct regions, such as separating tumor tissue from healthy tissue.

MITKats: Medical Imaging Interaction Toolkit with augmented tools for segmentation, an open-source set of segmentation tools presented here.

3   Background

Manual segmentation is typically performed with 2D tools to delineate edges on each image slice. While manual segmentation is the gold standard [19], it is too time-consuming for large analyses, sometimes taking upwards of an hour per image series [20]. Because of this, many fully automatic and semi-automatic algorithms have been created to expedite the segmentation process. Reviews of these methods can be found in [21,22]. Benchmarks of fully automatic algorithms have demonstrated encouraging accuracies [23], but acceptance of these methods in the clinic is limited due to concerns about errors and transparency [24]. Semi-automatic algorithms draw a balance by unifying the power of computer processing with the intuition of the human operator. However, existing semi-automatic programs still need improvement with regards to operator time [25] and user-friendliness [26].

To address these drawbacks, this work provides and validates an accessible semi-automatic protocol for the fast segmentation of glioblastomas. Specific goals included the creation of an intuitive computer-assisted segmentation utility, addition of 3D editing tools for manual correction, and integration within a user-friendly environment. These aims were implemented in the Medical Imaging Interaction Toolkit (MITK) as an extension of its existing Segmentation plugin [27,28]. This modified software, MITK with augmented tools for segmentation (MITKats), is also a free and open-source program. By addressing the aforementioned goals with MITKats, we expedite semi-automatic segmentation by introducing validated, easy-to-use 3D tools.

4   Methods

4.1   Data Sources

A total of 38 3D MRIs of brain tumors were obtained from two publicly available datasets. 20 cases of high-grade gliomas (including glioblastomas and anaplastic astrocytomas) were obtained from the Multimodal Brain Tumor Segmentation Challenge (BRATS) 2012 [23], a notable accuracy benchmark. Four modalities were available for the BRATS dataset: T1, T1 with contrast (T1c), T2, and T2 FLAIR, though MITKats only used T1c and FLAIR images. The ground truths were derived from all four modalities and designated into 3 regions:

- Active: The contrast-enhancing parts of the tumor, seen on T1c
- Core: The Active component plus non-enhancing features seen on T1, including necrosis
- Whole: The hyper-intense region on T2 and FLAIR, corresponding to edema

18 cases of glioblastoma were obtained from a previously published study performed at St. Olav's University Hospital [25]. Only T1c MRIs were used to create segmentations, which were generated in BrainVoyager [29], 3D Slicer [30], and ITK-SNAP [31]. A single tumor region was labeled, comprising enhancing and necrotic regions. Given that the authors did not observe non-enhancing tumor regions, their definition of tumor was closest to that of the Core component from BRATS. While no ground truths were designated in this dataset, each segmentation was timed, serving as a speed benchmark.

Both datasets had originally been pre-processed, resulting in interpolation of image resolutions to a 1 mm isotropic voxel size. Images from BRATS had been skull-stripped, while those from St. Olav's had not.

4.2   Algorithmic Development

Two modifications were made to the MITK framework in order to expedite the segmentation process in MITKats. First, the Threshold Components tool was added, which expanded connected threshold segmentation to accept multiple seed points (and thus separated regions). It also allowed for the manipulation of seed points independently of the thresholds, as well as streamlining out unnecessary user input. The second modification was adding in segmentation capability to the Clipping Plane View, which was originally intended for volume measurements only. Therefore, a segmentation could be graphically adjusted in three dimensions through extraction of a clipped piece.

MITKats can be found here: https://github.com/RabadanLab/MITKats, and is in the process of being merged onto the main branch of MITK.

4.3   Segmentation Protocol

Segmentations were performed in MITKats by A.X.C., a medical student who had used similar software to segment 93 glioblastoma cases as part of a previous project [32]. Segmentation time, as measured by stopwatch, was started after loading the original image(s) and stopped after opening the save segmentation dialog, thus ignoring the time for file operations.

Thresholding. A new segmentation label was created for the T1c image, and the Segmentation plugin view is opened. Selecting the newly implemented Threshold Components tool, seed point(s) were placed in enhancing region(s), and the lower threshold adjusted until the apparent hyper-intensities were all included (Fig. 1a). If the tumor was grossly non-enhancing, another label was created where seed point(s) were placed in hypo-intense regions, and the upper threshold adjusted.

Cropping. Regions of normal brain tissue that were erroneously included as part of the Thresholding process were removed in two ways. Precise exclusion of regions was done via the Clipping Plane tool, where up to 6 deformable 3D surfaces were superimposed on the segmentation. This allowed the generalized separation of erroneous regions from the tumor body (Fig. 1b), a new feature added in MITKats. For gross corrections, the existing Image Cropping tool could be used instead. A 3D rectangular bounding box was graphically defined and used to mask and overwrite the original segmentation. Finally, if the true tumor contained only one connected component, either of these methods could be followed up with the Picking tool to exclude erroneous regions that were severed from the tumor.

Smoothing and Filling. The segmentation was smoothed via the Closing tool, a part of the existing set of Morphological Operations, typically with a radius of 2. This label was saved as the Active component. For the Core volume, the non-enhancing label could be joined to this component via the Union tool, if applicable. To include areas of necrosis, the Closing and Fill Holes tools were used (both typically with radius of 10), and this new label was saved as the Core component (Fig. 1c).

Whole Tumor. For segmenting the Whole tumor component of the BRATS dataset, the above protocol for obtaining the Core region was repeated for the FLAIR image.

4.4 Accuracy and Speed Analysis

The accuracy of MITKats segmentations was assessed by comparison to the reference segmentations provided by BRATS and St. Olav's datasets. In BRATS, the references were ground truths fused from 4 expert manual annotations. Each of the Whole, Core, and Active components were compared to their reference counterparts via Dice score [33] as well as calculated tumor volume. The mean inter-rater Dice scores from BRATS were used as controls. This was because

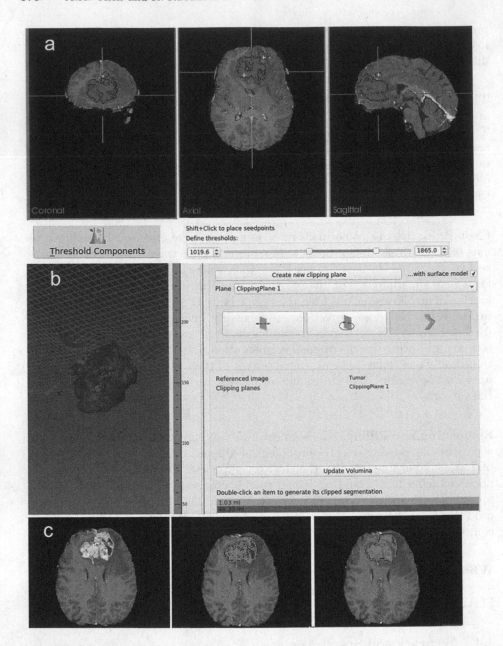

Fig. 1. Demonstration of a user-friendly semiautomatic segmentation protocol. (a) The Threshold Components tool allows the user to set an intensity threshold and multiple seedpoints (yellow crosses). Regions within the threshold and connected to the seedpoints are selected. (b) The deformable clipping plane allows 3D correction of the segmentation, typically useful for removing leaked regions. (c) Original image, clipped image, and final morphological operations such as Closing and Fill Holes smooth the enhancing segmentation into a Core segmentation. (Color figure online)

the fused ground truths were derived from the individual expert annotations, artificially bolstering the Dice score between any given expert and the fused standard.

For comparing data in the St. Olav's cohort, the MITKats segmentations were assessed via Dice score and volume to all 12 reference segmentations (3 softwares × 2 raters × 2 repetitions). As a control, Dice scores were calculated only for pairs of reference segmentations created by different raters. The time to create the Core component of the MITKats segmentation was compared against the times reported in the dataset.

5   Results

Aggregate Dice scores and volume estimates of MITKats segmentations with respect to BRATS ground truths for each tumor region are shown in Table 1. The Dice similarity of MITKats segmentations compared to the ground truth was equivalent to inter-rater variability for the Whole and Active tumor regions, but was worse for the Core region. Volume measurements averaged over all regions were 94% of those estimated by the ground truth, with a mean fractional error of 18%.

Table 1. Segmentation accuracy approaches inter-rater agreement in the Brain Tumor Segmentation Challenge. Twenty high-grade glioma cases were segmented using MITKats and compared against ground truths from BRATS. Three regions (Whole, Core, and Active) were segmented for each patient, and the mean Dice scores (± standard deviation) are shown. The volumes of each region were also compared to the ground truth.

Tumor compartment		Whole	Core	Active
Dice score (%)	MITKats v. BRATS	88 ± 5	84 ± 10	77 ± 14
	BRATS Inter-rater	88 ± 2	93 ± 3	74 ± 13
Volume ratio		0.96	1.06	0.81
Volume relative error		10%	25%	20%

A comparison of estimated volumes for each case across both datasets is shown in Fig. 2, including a logarithmic Bland-Altman analysis. While the volumes segmented by MITKats were similar to reference segmentations, they tended to be underestimated, particularly for the Active tumor region of the BRATS dataset.

The Core component of the MITKats segmentation was pairwise compared to segmentations performed by other softwares in the St. Olav dataset. The average Dice scores of MITKats segmentations compared to each of the reference segmentations was 0.88, compared against their inter-rater agreement of 0.94 (Fig. 3a).

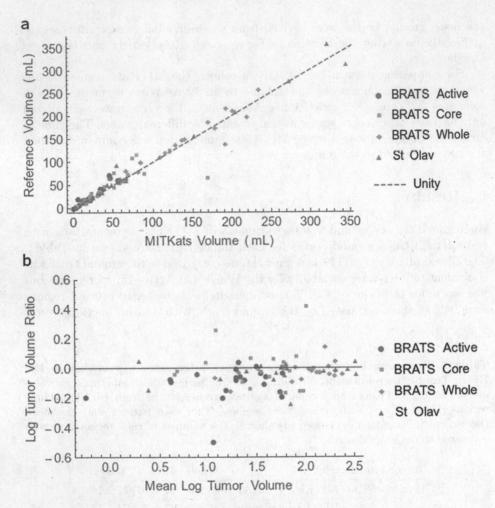

Fig. 2. Comparison of segmented volumes by tumor region. (a) Volumes calculated using MITKats are compared against their reference standards for each tumor component. (St. Olav segmentations designated only a single Core-like component.) (b) A logarithmic Bland-Altman plot is shown for different regions of segmentations, comparing the ratio of segmented volumes to the averages of the base 10 logarithms of tumor volume (in mL). There is a tendency for MITKats to underestimate tumor volume at low sizes, particularly in the Active tumor component.

The time for segmentation is also compared to the St. Olav's dataset in Fig. 3b. The speed of MITKats as compared to those reported was an average of 4, 5, and 11 times faster than ITK-SNAP, 3D Slicer, and BrainVoyager, respectively. The typical Core segmentation using MITKats was 4.2 ± 2.0 min on the St. Olav's dataset and 4.0 ± 3.1 min on the BRATS dataset, where uncertainties represent standard deviation.

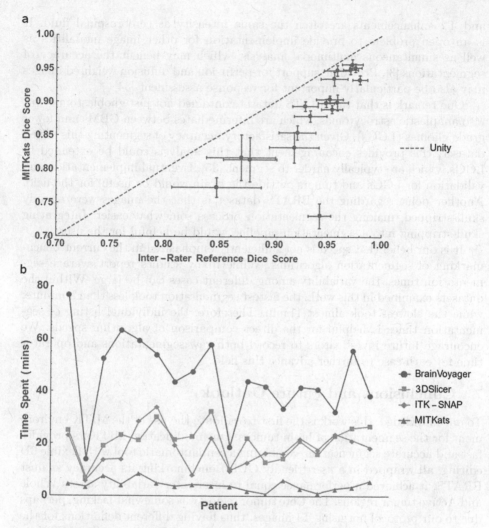

Fig. 3. MITKats is faster than other segmentation softwares, while approaching optimal accuracy. (a) Eighteen glioblastoma cases were segmented using MITKats and compared against the segmentations performed in [25]. Points represent the average Dice score when compared to all other segmentations performed by different raters. Error bars represent standard deviation. (b) The time required for segmenting via MITKats is compared to those using BrainVoyager, 3DSlicer, and ITK-SNAP (mean of 2 trials each).

6 Open Problems

The scope of this work has some limitations for which we hope future studies will address. In terms of protocol and datasets, this pipeline works best for T1c and FLAIR images, because T1 without contrast has limited enhancement, and T2 enhancements are often the same intensity as cerebrospinal fluid. It is an open problem to provide implementation for other image modalities as well as simultaneous multimodal analysis, which may benefit the accuracy of segmentations [3]. Providing support for perfusion and diffusion weighted images may also be particularly important for response assessment [34].

One remark is that the BRATS dataset contained not just glioblastomas but also anaplastic astrocytomas, which are intermediates between GBMs and lower grade gliomas (LGGs). Given the satisfactory accuracy of segmenting this mixed dataset, this provides encouragement that this analysis could be extended to LGGs, which are typically harder to segment [35]. Eventual implementation and validation for LGGs and tumors outside the brain would be useful for the field. Another point regarding the BRATS dataset is that the images were already skull-stripped, making the segmentation process somewhat easier. Integrating skull-stripping into a segmentation pipeline would be helpful for the user.

It is our belief that speed is not sufficiently emphasized in the current benchmarking of segmentation algorithms. While many studies report average segmentation times, the variability among different cases can be large. Within the datasets examined in this work, the fastest segmentation took less than a minute, while the slowest took almost 11 min. Therefore, the individual listing of segmentation times is helpful for the direct comparison of algorithm speeds. We encourage future investigators to record both raw segmentations and operator time for each case to further advance this field.

7   Conclusions and Future Outlook

To our knowledge, this work is the first to validate the use of the MITK environment for the segmentation of brain tumors. Our modification MITKats provides fast and accurate segmentation, combining a semiautomatic tool with flexible 3D editing, all wrapped in a user-friendly GUI. Benchmarking its accuracy against BRATS, it achieved a performance equal to inter-rater variability across Whole and Active tumor regions. The Core tumor region was somewhat lacking, perhaps due to our protocol not using T1 images, thus having different definitions for the component. Benchmarking its speed against the St. Olav's dataset, MITKats was over 4 times faster than its quickest competitor. While its accuracy was somewhat lower than inter-rater variability, the reference segmentations were also semi-automatically derived, and therefore not necessarily the ground truth. Finally, MITK is an open source toolkit which encourages the free use and continued development of this software.

The segmentations produced by MITKats form a starting point for a variety of data analysis techniques. Texture features [36], morphological characteristics [13], and topological summaries [32] can be extracted from the delineations and used to classify the cancer into spatial phenotypes. They can also be used to aid volumetric assessments [8] and validate growth models [37,38]. The quantified image features obtained from these segmentations, alongside the abundance of sequencing and histological data [39], will effectively support the further exploration of the radiogenomics of glioblastoma.

Acknowledgments. We would like to thank Anthea Monod and the rest of the Rabadan lab for their helpful comments and feedback.

References

1. DeAngelis, L.M.: Brain tumors. New Engl. J. Med. **344**(2), 114–123 (2001)
2. Mabray, M.C., Barajas, R.F., Cha, S.: Modern brain tumor imaging. Brain Tumor Res. Treat. **3**(1), 8–23 (2015)
3. Dupont, C., Betrouni, N., Reyns, N., Vermandel, M.: On image segmentation methods applied to glioblastoma: state of art and new trends. IRBM **1**(3), 1–13 (2016)
4. Bauer, S., Lu, H., May, C.P., Nolte, L.P., Büchler, P., Reyes, M.: Integrated segmentation of brain tumor images for radiotherapy and neurosurgery. Int. J. Imaging Syst. Technol. **23**(1), 59–63 (2013)
5. Kickingereder, P., Burth, S., Wick, A., et al.: Radiomic profiling of glioblastoma: identifying an imaging predictor of patient survival with improved performance over established clinical and radiologic risk models. Radiology **280**(3), 880–889 (2016)
6. Cui, Y., Tha, K.K., Terasaka, S., Yamaguchi, S., Wang, J., Kudo, K., Xing, L., Shirato, H., Li, R.: Prognostic imaging biomarkers in glioblastoma: development and independent validation on the basis of multiregion and quantitative analysis of MR images. Radiology **287**, 546–553 (2015)
7. Chow, D.S., Qi, J., Guo, X., Miloushev, V.Z., Iwamoto, F.M., Bruce, J.N., Lassman, A.B., Schwartz, L.H., Lignelli, A., Zhao, B., Filippi, C.G.: Semiautomated volumetric measurement on postcontrast MR imaging for analysis of recurrent and residual disease in glioblastoma multiforme. Am. J. Neuroradiol. **35**(3), 498–503 (2014)
8. Clarke, L.P., Velthuizen, R.P., Clark, M., Gaviria, J., Hall, L., Goldgof, D., Murtagh, R., Phuphanich, S., Brem, S.: MRI measurement of brain tumor response: comparison of visual metric and automatic segmentation. Magn. Reson. Imaging **16**(3), 271–279 (1998)
9. Yang, D., Rao, G., Martinez, J., Veeraraghavan, A., Rao, A.: Evaluation of tumor-derived MRI-texture features for discrimination of molecular subtypes and prediction of 12-month survival status in glioblastoma. Med. Phys. **42**(11), 6725 (2015)
10. Hu, L.S., Ning, S., Eschbacher, J.M., et al.: Radiogenomics to characterize regional genetic heterogeneity in glioblastoma. Neuro Oncol. **19**, 135 (2016)
11. Kickingereder, P., Bonekamp, D., Nowosielski, M., et al.: Radiogenomics of glioblastoma: machine learning based classification of molecular characteristics by using multiparametric and mutiregional MR imaging features. Radiology **281**, 907–918 (2016)
12. Czarnek, N., Clark, K., Peters, K.B., Mazurowski, M.A.: Algorithmic three-dimensional analysis of tumor shape in MRI improves prognosis of survival in glioblastoma: a multi-institutional study. J. Neuro Oncol. **132**(1), 55–62 (2017)
13. Mazurowski, M.A., Zhang, J., Peters, K.B., Hobbs, H.: Computer-extracted MR imaging features are associated with survival in glioblastoma patients. J. Neuro Oncol. **120**(3), 483–488 (2014)
14. Itakura, H., Achrol, A.S., et al.: Magnetic resonance image features identify glioblastoma phenotypic subtypes with distinct molecular pathway activities. Sci. Transl. Med. **7**(303), 303ra138 (2015)

15. Jeanquartier, F., Jean-Quartier, C., Kotlyar, M., Tokar, T., Hauschild, A.C., Jurisica, I., Holzinger, A.: Machine Learning for In Silico Modeling of Tumor Growth BT - Machine Learning for Health Informatics: State-of-the-Art and Future Challenges, pp. 415–434. Springer International Publishing, Cham (2016)
16. Macyszyn, L., Akbari, H., Pisapia, J.M., Da, X., Attiah, M., et al.: Imaging patterns predict patient survival and molecular subtype in glioblastoma via machine learning techniques. Neuro Oncol. 18(3), 417–425 (2016)
17. Jeanquartier, F., Jean-Quartier, C., Schreck, T., Cemernek, D., Holzinger, A.: Integrating open data on cancer in support to tumor growth analysis. In: Renda, M.E., Bursa, M., Holzinger, A., Khuri, S. (eds.) ITBAM 2016. LNCS, vol. 9832, pp. 49–66. Springer, Cham (2016). doi:10.1007/978-3-319-43949-5_4
18. Clark, K., Vendt, B., Smith, K., Freymann, J., Kirby, J., Koppel, P., Moore, S., Phillips, S., Maffitt, D., Pringle, M., Tarbox, L., Prior, F.: The cancer imaging archive (TCIA): maintaining and operating a public information repository. J. Digit. Imaging 26(6), 1045–1057 (2013)
19. Porz, N., Bauer, S., Pica, A., Schucht, P., Beck, J., Verma, R.K., Slotboom, J., Reyes, M., Wiest, R.: Multi-modal glioblastoma segmentation: man versus machine. PLoS ONE 9(5), 1–9 (2014)
20. Kaus, M.R., Warfield, S.K., Nabavi, A., Black, P.M., Jolesz, F.A., Kikinis, R.: Automated segmentation of MR images of brain tumors. Radiology 218(2), 586–591 (2001)
21. Bauer, S., Wiest, R., Nolte, L.P., Reyes, M.: A survey of MRI-based medical image analysis for brain tumor studies. Phys. Med. Biol. 58(13), R97–R129 (2013)
22. Wang, J., Liu, T.: A survey of MRI-based brain tumor segmentation methods. Tsinghua Sci. Technol. 19(6), 578–595 (2014)
23. Menze, B.H., Jakab, A., Bauer, S., et al.: The multimodal brain tumor image segmentation benchmark (BRATS). IEEE Trans. Med. Imaging 34(10), 1993–2024 (2015)
24. Gordillo, N., Montseny, E., Sobrevilla, P.: State of the art survey on MRI brain tumor segmentation. Magn. Reson. Imaging 31(8), 1426–1438 (2013)
25. Fyllingen, E.H., Stensjoen, A.L., Berntsen, E.M., Solheim, O., Reinertsen, I.: Glioblastoma segmentation: comparison of three different software packages. PLoS ONE 11(10), e0164891 (2016)
26. Ramkumar, A., Dolz, J., Kirisli, H.A., et al.: User interaction in semi-automatic segmentation of organs at risk: a case study in radiotherapy. J. Digit. Imaging 29(2), 264–277 (2016)
27. Wolf, I., Vetter, M., Wegner, I., Bottger, T., Nolden, M., Schobinger, M., Hastenteufel, M., Kunert, T., Meinzer, H.P.: The medical imaging interaction toolkit. Med. Image Anal. 9(6), 594–604 (2005)
28. Maleike, D., Nolden, M., Meinzer, H.P., Wolf, I.: Interactive segmentation framework of the Medical Imaging Interaction Toolkit. Comput. Methods Prog. Biomed. 96(1), 72–83 (2009)
29. Goebel, R., Esposito, F., Formisano, E.: Analysis of functional image analysis contest (FIAC) data with BrainVoyager QX: from single-subject to cortically aligned group general linear model analysis and self-organizing group independent component analysis. Hum. Brain Mapp. 27(5), 392–401 (2006)
30. Egger, J., Kapur, T., Fedorov, A., Pieper, S., Miller, J.V., Veeraraghavan, H., Freisleben, B., Golby, A.J., Nimsky, C., Kikinis, R.: GBM volumetry using the 3D Slicer medical image computing platform. Sci. Rep. 3, 1364 (2013)

31. Yushkevich, P.A., Piven, J., Hazlett, H.C., Smith, R.G., Ho, S., Gee, J.C., Gerig, G.: User-guided 3D active contour segmentation of anatomical structures: significantly improved efficiency and reliability. NeuroImage **31**(3), 1116–1128 (2006)
32. Crawford, L., Monod, A., Chen, A.X., Mukherjee, S., Rabadán, R.: Topological Summaries of Tumor Images Improve Prediction of Disease Free Survival in Glioblastoma Multiforme. Arxiv pre-print (Nov)
33. Dice, L.R.: Measures of the amount of ecologic association between species. Ecology **26**(3), 297–302 (1945)
34. Huang, R.Y., Neagu, M.R., Reardon, D.A., Wen, P.Y.: Pitfalls in the neuroimaging of glioblastoma in the era of antiangiogenic and immuno/targeted therapy - detecting illusive disease, defining response. Front. Neurol. **6**, 1–16 (2015)
35. Akkus, Z., Sedlar, J., Coufalova, L., Korfiatis, P., Kline, T.L., Warner, J.D., Agrawal, J., Erickson, B.J.: Semi-automated segmentation of pre-operative low grade gliomas in magnetic resonance imaging. Cancer Imaging **15**(1), 1–10 (2015)
36. Upadhaya, T., Morvan, Y., Stindel, E., Le Reste, P.J., Hatt, M.: Prognosis classification in glioblastoma multiforme using multimodal MRI derived heterogeneity textural features: impact of pre-processing choices, vol. 9785, 97850W, March 2016
37. Jeanquartier, F., Jean-Quartier, C., Cemernek, D., Holzinger, A.: In silico modeling for tumor growth visualization. BMC Syst. Biol. **10**(1), 59 (2016)
38. Jean-Quartier, C., Jeanquartier, F., Cemernek, D., Holzinger, A.: Tumor growth simulation profiling. In: Renda, M.E., Bursa, M., Holzinger, A., Khuri, S. (eds.) ITBAM 2016. LNCS, vol. 9832, pp. 208–213. Springer, Cham (2016). doi:10.1007/978-3-319-43949-5_16
39. Cooper, L.A.D., Kong, J., Gutman, D.A., Wang, F., Gao, J., Appin, C., Cholleti, S., Pan, T., Sharma, A., Scarpace, L., Mikkelsen, T., Kurc, T., Moreno, C.S., Brat, D.J., Saltz, J.H.: Integrated morphologic analysis for the identification and characterization of disease subtypes. J. Am. Med. Inf. Assoc. JAMIA **19**(2), 317–323 (2012)

Topological Characteristics of Oil and Gas Reservoirs and Their Applications

V.A. Baikov[1], R.R. Gilmanov[2], I.A. Taimanov[3,4(✉)], and A.A. Yakovlev[2]

[1] Ufa State Aviation Technical University, 450025 Ufa, Russia
baikov@ufanipi.ru
[2] OOO "Gazpromneft NTC", 190000 St. Petersburg, Russia
{Gilmanov.RR,Yakovlev.AAle}@gazpromneft-ntc.ru
[3] Sobolev Institute of Mathematics, 630090 Novosibirsk, Russia
[4] Novosibirsk State University, 630090 Novosibirsk, Russia
taimanov@math.nsc.ru

Abstract. We demonstrate applications of topological characteristics of oil and gas reservoirs considered as three-dimensional bodies to geological modeling.

Keywords: Geological modeling · Oil and gas reservoirs · Betti numbers · Euler characteristic · Persistent homology · Bottleneck distance

At present no company develops oil and gas fields without constructing geologic and hydrodynamic models. This is due in particular to the fact that recently the emphasis in design, planning and monitoring has shifted to over-dissected and low-permeability reservoirs. To assess economic efficiency and optimal placement of wells and to predict hydrocarbon production levels, it is important to have some quantitative representation of the object under study. This requires a mathematical measure of a geological description and mathematical models of the structure of oil and gas reservoirs.

The geological modeling based on digital oil and gas reservoirs splits into two parts:

1. a digital interpolation of a reservoir based on the observed data and on the probabilistic nature of functions which describe formations;
2. a hydrodynamic modeling based on the filtration equations.

Therewith it is important to choose the most suitable model for developing. In [1] we proposed to use topological characteristics of digital reservoirs as one of the factors for choosing a model. These characteristics can be used for

- *comparing different stochastic realizations of the same reservoir and, in particular, using that information for choosing a certain realization for industrial development;*
- *estimating the topological complexity of a reservoir.*

A. Holzinger et al. (Eds.): Integrative Machine Learning, LNAI 10344, pp. 182–193, 2017.
https://doi.org/10.1007/978-3-319-69775-8_11

In particular, this method can help to choose realization that gives a reliable model of a reservoir and whose exploration does not need resource-intensive calculations.

For optimizing a process of geological and hydrodynamic modeling by limiting a series of direct problems there arise tasks of creating a list of topological, geometric, fractal, and other characteristics of inhomogeneous anisotropic environment and their subsequent influences on the construction of a model. Such problems are studied in geometry of random fields.

In [1] we demonstrated that two different stochastic approaches for constructing digital reservoirs gives topologically similar pictures and one of them, being more rough is nevertheless preferable for dynamical modeling due to its relative simplicity for numerical dynamical modeling. We expose some results on the Betti numbers of reservoirs in Sect. 2.

Since the "permeability" function Z determines a natural filtration of the reservoir by the excursion sets it is reasonable to pick up the topological picture of the filtration and use for that the persistent homology [2,3] (see also [4–7]). In this framework

- the "bottleneck" distance between persistent diagrams can be used for estimating differences between reservoirs and not only between their models.

We discuss this approach in Sect. 3.

1 Stochastic and Topological Preliminaries

1.1 The Kriging

The digital reservoirs under consideration are constructed by the kriging method from the observed data (see [8–10] and the references therein). This method has many variations, based on the same idea, and we explain which one we use.

In our case a digital reservoir is a union of cubes such that a certain characteristic related to the permeability is a function Z on the set of these cubes. For simplicity we assume that the reservoir is the domain

$$D = \{x_0 \leq x \leq x_0 + N_x \delta_x, y_0 \leq y \leq y_0 + N_y \delta_y, z_0 \leq z \leq z_0 + N_z \delta_z\},$$

where x and y are the lateral coordinates and z is the height coordinate, while $\delta_x, \delta_y, \delta_z$ are the length, width and depth of the elementary cube. The domain D splits into $N_x N_y N_z$ elementary cubes C_{k_x, k_y, k_z} defined by the inequalities

$$x_0 + (k_x - 1)\delta_x \leq x \leq x_0 + k_x \delta_x, \quad y_0 + (k_y - 1)\delta_y \leq y \leq y_0 + k_y \delta_y,$$

$$z_0 + (k_z - 1)\delta_z \leq z \leq z_0 + k_z \delta_z,$$

where the triples (k_x, k_y, k_z) parameterize the elementary cubes and Z is considered as a function on these triples:

$$Z = Z(k_x, k_y, k_z).$$

We denote the set of all these triples by

$$S = \{1, \ldots, N_x\} \times \{1, \ldots, N_y\} \times \{1, \ldots, N_z\}$$

and for every subset $S' \subset S$ we denote by $D_{S'}$ the union of elementary cubes corresponding to triples from S':

$$D_{S'} \subset D, \quad D_S = D.$$

Let the function Z is known for some set S' of elementary cubes. For instance, this may be the observed data from wells. We extend Z onto S by the following stochastic regression method.

Let us choose a procedure for choosing randomly an element p_{M+1} from $S \backslash S'$ where M is the number of elements of S'.

The function Z is considered as a random field such that

1. it is stationary, i.e., it has the same expectations at all points:

$$\mathrm{E}\,(Z(p)) = \mathrm{E}\,(Z(q)) = m \quad \text{for all } p, q \in S$$

and m is known (*simple kriging*);

2. the correlation between two random variables depends only on the spatial distance between them:

$$C(Z(p), Z(q)) = C(|p - q|).$$

Here we mean by the spatial distance $|p - q|$ between elementary cubes the distance between their centers. The correlators are given by the variogram:

$$\gamma(|h|) = \frac{1}{2}\mathrm{E}\,((Z(p) - Z(p + h))^2) = C(0) - C(|h|).$$

This variogram is derived from observations.

Given a sample $(Z(p_1), \ldots, Z(p_M))$, the values of Z at S', the value s_{M+1} is obtained from the conditions

$$Z^* = \sum_{i=1}^{M} \lambda_i Z(p_i), \quad \sum \lambda_i = 1, \quad \mathrm{E}\,((s_{M+1} - Z^*)^2) \to \min$$

which results in the system:

$$\sum \lambda_i = 1,$$

$$\sigma^2 = C(0) - 2\sum_i \lambda_i C(|p_i - p_0|) + \sum_{i,j} \lambda_i \lambda_j C(|p_i - p_j|) \to \min,$$

where σ^2 is a measure of precision. To determine $Z(p_{M+1})$ we put

$$Z(p_{M+1}) = s_{M+1} + \xi$$

where the random process ξ satisfies the Gauss distribution with $E = 0, \text{Var} = \sigma^2$. Thus we derive the new sample $Z(p_1), \ldots, Z(p_{M+1})$, add p_{M+1} to S' and resume by the same way until we extend Z onto S.

This procedure is called *the sequential Gauss simulation* (SGS). The standard variograms that are used are

- the Gaussian variogram: $C(h) = \text{const}\, (1 - e^{-h^2/R^2})$,
- the exponential variogram: $C(h) = \text{const}\, (1 - e^{-h/R})$.

In both cases R is the radius of a variogram.

There is another stochastic regression in which the field is represented as a linear combination of the first M Legendre polynomials

$$Z(x, y, h) = \sum_{i=1}^{M} a_i(x, y) L_i(h)$$

where x and y are the lateral variables, h is the depth, and $a_i(x, y)$ are independent random fields which are extrapolated by some two-dimensional stochastic regression. This method is called the *spectral expansion*.

In developing oil formations an important data is

$$Z(p) = \text{GL}(p)$$

which is the gamma logging, i.e., the natural radioactivity of formation. We put

$$\alpha(p) = \frac{\text{GL}(p) - \text{GL}_{\min}}{\text{GL}_{\max} - \text{GL}_{\min}}$$

and assume that p belongs to the formation if

$$\alpha(p) \leq \alpha_0,$$

where α_0 is the excursion coefficient. By varying α_0 we obtain a filtration of the reservoir D by the excursion sets

$$D_1 \subset D_2 \subset \cdots \subset D_M \subset D_\infty = D$$

where $\varepsilon_1 < \varepsilon_2 < \cdots < \varepsilon_M < 1$ and $D_i = D_{S_{\varepsilon_i}}, S_{\varepsilon_i} = \{\alpha \leq \varepsilon_i\} \subset S$.

The double difference parameter α is widely used in practice. Therewith $\text{GL}_{\min}$ and $\text{GL}_{\max}$ are the minimal and maximal values of GL which correspond to a neat oil and gas reservoir and a clay which supports a reservoir. These values should not be confused with the absolute minimum and maxima values of GL with which they coincide only in exceptional model examples. This is due, in particular, to the possible presence of minor anomaly noises and to effects of stochastic modeling. Moreover, often $\text{GL}_{\min}$ and $\text{GL}_{\max}$ are calibrated by certain samples. Therefore sometimes in modeling there appear values of α which are less than zero or greater or equal than 1 and, if not to take care of that, D_∞ may not coincide with D.

The stochastic modeling leads to very adequate pictures of reservoirs. As an example, we present on Fig. 1 the model of a reservoir obtained from the observed data. This model is obtained by the SGS data, the parameters of the domain D are $N_x = N_y = 120$, $N_z = 490$, $\delta_x = \delta_y = 50\,\mathrm{m}$, $\delta_z = 0.4\,\mathrm{m}$, the colors vary from light to dark that corresponds to the variation of α from small to large values, the reservoir corresponds to the excursion $\alpha = 0.6$.

Fig. 1. A reservoir modeled by the SGS method

1.2   Topological Characteristics of 3-Dimensional Bodies

We consider three-dimensional solid bodies as composed from elementary cubes (cubic complexes).

The main topological characteristics of such bodies are their Betti numbers (with $\mathbb{Z}_2$ coefficients) b_0, b_1 and b_2. The meaning of these characteristics is very natural: b_0 is the number of connected components, b_1 is the number of handles, and b_2 is the number of holes (cavities).

If we start from a solid cube, remove k holes from its interior and attach l handles to the cube we obtain the body X for which $b_0 = 1, b_1 = l, b_2 = k$. The Betti numbers of a topological space are the ranks of the corresponding homology groups $H_i(X; \mathbb{Z}_2)$ (here an in the sequel we consider the homology groups with coefficients with $\mathbb{Z}_2$ and for denote them by $H_i(X)$ for simplicity):

$$H_0 = \mathbb{Z}_2^{b_0} = \mathbb{Z}_2 \oplus \cdots \oplus \mathbb{Z}_2 \text{ (the sum of } b_0 \text{ copies of } \mathbb{Z}_2),$$

$$H_1 = \mathbb{Z}_2^{b_1}, \quad H_2 = \mathbb{Z}_2^{b_2}.$$

The alternated sum

$$\chi = b_0 - b_1 + b_2$$

is called the Euler characteristic of a three-dimensional solid body.

We refer to [6] for an introductory exposition, of these topological characteristics, oriented to applications. We recall that if two topological spaces (bodies) are topologically equivalent (or homeomorphic), i.e. if there exists a continuous in both sides one-to-one correspondence between points of spaces, then they have the same topological characteristics (the Betti numbers, homology groups etc.)

The Euler characteristic may be easily computed from the cubic decomposition of the solid body X. We have X as a union of cubes such that

(a) two different cubes may intersect each other only by a joint vertex, edge or face;
(b) two different faces may intersect each other only by a joint vertex or edge;
(c) two different edges my intersect each other only by a joint vertex.

We denote by c_0 the number of vertices; by c_1 the number of edges; by c_2 the number of faces; and by c_3 the number of cubes. For instance, the cubic decomposition of an elementary cube has 8 vertices, 12 edges, 6 faces and 1 cube.

The Euler characteristic of a three-dimensional body is given also by the formula:

$$\chi = c_0 - c_1 + c_2 - c_3. \tag{1}$$

For an elementary cube we have $\chi = 8 - 12 + 6 - 1 = 1$.

If X is body composed from finitely many cubes and lies in the three-space $\mathbb{R}^3$, then the particular case of the Alexander duality implies that

$$b_2(X) = b_0(\mathbb{R}^3 \setminus X) - 1.$$

Hence, in difference with higher-dimensional space, for calculating the Betti numbers of a three-dimensional body X it is enough to find its Euler characteristic χ from the cubic decomposition (see (1)) and the numbers of connected components of X and of its complement. Then b_1 is given by the equality

$$b_1(X) = b_0(X) + b_0(\mathbb{R}^3 \setminus X) - 1 - \chi(X). \tag{2}$$

That drastically simplifies the calculation of the Betti numbers and reduces it to calculating of the numbers of connected components of cubic complexes.

The development of numerical methods for calculating the Betti numbers is necessary because reservoirs may be very complicated. In particular, some numerical approach, based on a certain discretization of the Morse theory to finding the Betti numbers of reservoirs was exposed in [11].

2   The Betti Numbers of Digital Reservoirs

Since geological formations are natural examples of three-dimensional solid bodies, it is reasonable to consider their topology for geological applications however that was started not long ago (see [1] for oil and gas reservoirs and [12,13] and references therein for applications to structural geology).

Let us demonstrate numerical examples of the Betti numbers of reservoirs.

Given the excursion parameter α_0, we have the cubic complex

$$D_{\alpha_0} = \cup_{Z(k_x, k_y, k_z) \leq \alpha_0} C_{k_x, k_y, k_z}$$

composed from all cubes for which $Z \leq \alpha_0$.

Table 1. The Betti numbers and the Euler characteristic

α_0	b_0	b_1	b_2	χ
0.1	1664	0	0	1664
	477	1	0	476
	5042	1	0	5041
	3491	2	0	3489
0.2	4751	9	0	4742
	1330	10	0	1320
	18691	9	0	18682
	11779	60	0	11719
0.3	6113	260	0	5853
	1606	110	0	1496
	32601	495	3	32109
	18757	997	12	17772
0.4	1932	3682	3	−1747
	487	1150	0	−663
	20905	9355	329	11879
	12813	9455	391	3749
0.5	245	11389	163	−10981
	55	2995	29	−2911
	4971	45256	4187	−36098
	3713	28695	3324	−21658
0.6	18	8523	1434	−7071
	1	1927	265	−1661
	528	53806	18129	−35149
	473	29870	11421	−17976
0.7	1	3133	4903	1771
	1	545	1045	501
	14	28988	29705	731
	31	15949	16832	914
0.8	1	721	4132	3412
	1	92	974	883
	1	6658	17563	10906
	3	4216	10220	6007
0.9	1	85	1488	1404
	1	6	389	384
	1	637	4798	4162
	1	608	3171	2564

To construct from D_{α_0} the topological model of the corresponding reservoir we have to keep in mind that if two cubes do have only a joint edge or a joint vertex then there is no percolation between them through the joint cell (edge or vertex). The percolation between two adjacent cubes is possible only through a joint two-dimensional face. Hence we have to unstack all such cubes and obtain an abstract cubic complex X_{α_0}. This complex is the right model that respects the percolation rules and can be chosen for industrial development.

To give an impression on the topological complexity of reservoirs we present results of some calculations corresponding to simulated reservoirs (see Table 1). We consider the four digital models that correspond to the exponential variogram $C(h) = (1 - e^{-h/R})$ with $R = 500\,\mathrm{m}$ and $R = 1000\,\mathrm{m}$ and to the Gaussian variogram $C(h) = (1 - e^{-h^2/R^2})$ with $R = 500\,\mathrm{m}$ and $R = 1000\,\mathrm{m}$. The data of the reservoirs are $N_x = N_y = N_z = 100, \delta_x = \delta_y = 100\,\mathrm{m}, \delta_z = 1\,\mathrm{m}$.

The numerical experiment shows the stability of the integral topological characteristics (the Betti numbers weighted by a volume) under stochastic modeling, sensitivity to the type and the rank of the variogram (see Figs. 2 and 3). The characteristics lie on similar cycles and in both cases the inner (smaller) cycle corresponds to the largest value of $R(= 1000)$. Thus these characteristics can serve as classifiers for assigning digital geological models to equivalent and as a consequence, they have important applied values for the determination of analogs in the modeling of poorly studied oil fields (the case of lack of information for a reliable distribution of reservoir properties).

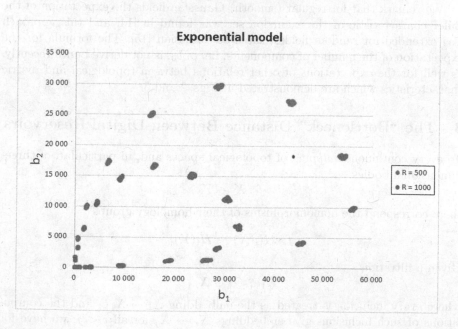

Fig. 2. Relations between the weighted Betti numbers of digital reservoirs for the exponential variogram

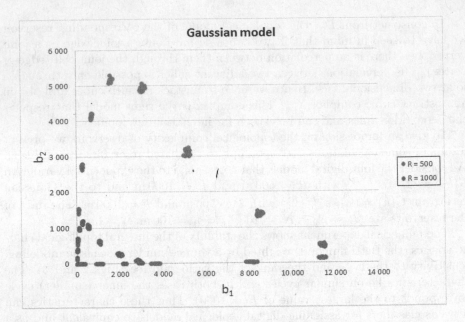

Fig. 3. Relations between the weighted Betti numbers of digital reservoirs for the Gaussianl variogram

We remark that for regular (smooth) Gaussian fields the expectation of the Euler characteristic of the excursion set was found in [14] and this approach was extended for random fields related to Gaussian [15]. The formula for the expectation of the number of components, i.e. of b_0, is not derived until recently, as well for the expectations of other relations between topological and metric characteristics which are demonstrated in Figs. 2 and 3.

3   The "Bottleneck" Distance Between Digital Reservoirs

To every continuous mapping of topological spaces and, in particular, of three-dimensional bodies

$$f : X \to Y$$

there correspond the homomorphisms of their homology groups

$$f_* : H_i(X) \to H_i(Y).$$

Given a filtration

$$X_1 \subset \cdots \subset X_M$$

where every inclusion is treated as the embedding $X_i \to X_{i+1}$ and the compositions of such inclusions give embeddings $X_i \to X_j$ for all $i < j$, we have for every dimension q the homomorphism

$$f_q^{i,j} : H_q(X_i) \to H_q(X_j).$$

The persistent homology groups [2–4] are defined as

$$H_q^{i,j} = \operatorname{Im} f_q^{i,j} = f_*(H_q(X_i)) \subset H_q(X_j).$$

Let us fix q. To every generator $z \in H_q(X_i)$ such that z does not lie in the image of $H_q(X_{i-1}) \to H_q(X_i)$, it is mapped into nontrivial elements by homomorphisms $H_q(X_i) \to H_q(X_{j-1})$ and $f_q^{i,j}(z) = 0$ we correspond a point on the plane with coordinates (i, j). Here we recall that we consider homology groups with coefficients in $\mathbb{Z}_2$ and this procedure is defined for all coefficients and also for continuous values of indices i.

The persistent diagram of a filtration (for the q-dimensional homology) is the union U of all such points taken with their multiplicities and points of of the diagonal $(x, x) \subset \mathbb{R}^2$ taken with infinite multiplicities.

The persistent homology and their persistent diagrams play a fundamental role in the modern topological data analysis [5, 7].

The persistent diagrams are stable under small perturbations of initial topological data [16]. The distance between different persistent diagrams is given by the bottleneck distance defined as follows

$$\rho(U, V) = \inf_{\eta:U \to V} \sup_{u \in U} |u - \eta(u)|,$$

where the infimum is taken over all bijections $\eta : U \to V$ and the norm $|u|$ on the plane has the form $|u| = |x| + |y|$ where $u = (x, y)$. This bottleneck distance plays an important role in the optimization theory and different algorithms for its computation were recently used in topological data analysis (see, for instance, [17, 18]).

The bottleneck distance can be used for comparing different digital reservoirs. We present results of some numerical experiments. We compute the bottleneck distances between the 0-dimensional $(q = 0)$ persistent diagrams corresponding to 8 digital reservoirs which splits into four pairs corresponding to the exponential (E) and Gaussian (G) variograms and to $R = 500\,\mathrm{m}$ or $R = 1000\,\mathrm{m}$. This reservoirs correspond to $N_x = N_y = N_z = 25$, $\delta_x = \delta_y = 400\,\mathrm{m}, \delta_z = 4\,\mathrm{m}$, and

Table 2. The bottleneck distance

	E500-1	E500-2	E1000-1	E1000-2	G500-1	G500-2	G1000-1	G1000-2
E500-1	0	0.11	0.11	0.12	0.13	0.09	0.15	0.16
E500-2	0.11	0	0.055	0.1	0.07	0.06	0.11	0.13
E1000-1	0.11	0.055	0	0.09	0.05	0.06	0.11	0.11
E1000-2	0.12	0.1	0.09	0	0.05	0.05	0.07	0.07
G500-1	0.13	0.07	0.05	0.05	0	0.07	0.08	0.08
G500-2	0.09	0.06	0.06	0.05	0.07	0	0.08	0.09
G1000-1	0.15	0.11	0.11	0.07	0.08	0.08	0	0.05
G1000-2	0.16	0.13	0.11	0.07	0.08	0.09	0.05	0

the step of the discretized excursion parameter α_0 (the step of the excursion filtration) is equal to $\Delta\alpha_0 = 0.01$. Metrically these reservoirs have the same form — $10000\,\text{m} \times 10000\,\text{m} \times 400\,\text{m}$ — as the reservoirs in Table 1. But we consider a rough decomposition because the complexity of the calculation of the bottleneck distance is $O(n^2 \log n)$ where n is the number of points in the persistence diagram and for some digital reservoirs from Table 1 we have $n \approx 50000$ which makes the calculation time- and resource-consuming. Keeping in mind that $0 \leq \alpha \leq 1$ and hence the distance between such diagrams is at most 1, the data shows that this metric really distinguishes diagrams but it needs to understand for which types of digital reservoirs and, in particular, for which ratios of R and the sizes of elementary cubes this approach gives applicable answers (Table 2).

References

1. Bazaikin, Y.V., Baikov, V.A., Taimanov, I.A., Yakovlev, A.A.: Numerical analysis of topological characteristics of three-dimensional geological models of oil and gas fields. Math. Model. (Matematicheskoe Modelirovanie) **25**(10), 19–31 (2013). (Russian)
2. Edelsbrunner, H., Letscher, D., Zomorodian, A.: Topological persistence and simplification. Discrete Comput. Geom. **28**, 511–533 (2002). doi:10.1007/s00454-002-2885-2
3. Zomorodian, A., Carlsson, G.: Computing persistent homology. Discrete Comput. Geom. **33**, 249–274 (2005). doi:10.1007/s00454-004-1146-y
4. Edelsbrunner, H., Harer, J.: Persistent homology – a survey. In: Goodman, J.E., Pach, J., Pollack R. (eds.) Surveys on Discrete and Computational Geometry. Contemporary Mathematics, vol. 453, pp. 257–282. American Mathematical Society, Providence, RI (2008)
5. Carlsson, G.: Topology and data. Bull. Am. Math. Soc. (N.S.) **46**, 255–308 (2009). doi:10.1090/S0273-0979-09-01249-X
6. Edelsbrunner, H., Harer, J.L.: Computational Topology: An Introduction. American Mathematical Society, Providence (2010)
7. Ferri, M.: Persistent topology for natural data analysis - A survey. https://arxiv.org/abs/1706.00411
8. Matheron, G.: Traité de Geostatistique Appliquée. Editions BGRM, Paris (1962)
9. Dubrule, O.: Geostatistics in Petroleum Geology. American Association of Petroleum Geologists, Tulsa (1998)
10. Baikov, V.A., Bakirov, N.K., Yakovlev, A.A.: Mathematical Geology. I. Introduction to Geostatistics. Izhevsk Institute of Computer Sciences, Izhevsk (2012). (Russian)
11. Bazaikin, Y.V., Taimanov, I.A.: On a numerical algorithm for computing topological characteristics of three-dimensional bodies. J. Comput. Math. Math. Phys. (Zhurnal Vychislitel'noi Matematiki i Matematicheskoi Fiziki) **53**, 523–530 (2013). (Russian)
12. Thiele, S.T., Jessel, M.W., Lindsay, M., Ogarko, V., Wellmann, J.F., Pakyuz-Charrier, E.: The topology of geology 1: Topological analysis. J. Struct. Geol. **91**, 27–38 (2016). doi:10.1016/j.jsg.2016.08.009

13. Thiele, S.T., Jessel, M.W., Lindsay, M., Wellmann, J.F., Pakyuz-Charrier, E.: The topology of geology 2: Topological uncertainty. J. Struct. Geol. **91**, 74–87 (2016). doi:10.1016/j.jsg.2016.08.010
14. Adler, R.J.: The Geometry of Random Fields. Wiley, London (1981)
15. Adler, R.J., Taylor, J.E.: Random Fields and Geometry. Springer, Heidelberg (2007)
16. Cohen-Steiner, D., Edelsbrunner, H., Harer, J.: Stability of persistence diagrams. Discrete Comput. Geom. **37**, 103–120 (2007). doi:10.1007/s00454-006-1276-5
17. Efrat, A., Itai, A., Katz, M.J.: Geometry helps in bottleneck matching and related problems. Algorithmica **31**(1), 1–28 (2001). doi:10.1007/s00453-001-0016-8
18. Kerber, M., Morozov, D., Nigmetov, A.: Geometry helps to compare persistence diagrams. https://arxiv.org/abs/1606.03357

Convolutional and Recurrent Neural Networks for Activity Recognition in Smart Environment

Deepika Singh[1(✉)], Erinc Merdivan[1,4], Sten Hanke[1], Johannes Kropf[1], Matthieu Geist[2,3,4], and Andreas Holzinger[5]

[1] AIT Austrian Institute of Technology, Wiener Neustadt, Austria
{deepika.singh,erinc.merdivan}@ait.ac.at
[2] Université de Lorraine, LORIA, UMR 7503, 54506 Vandoeuvre-lès-Nancy, France
[3] CNRS, LORIA, UMR 7503, 54506 Vandoeuvre-lès-Nancy, France
[4] LORIA, CentraleSupélec, Université Paris-Saclay, 57070 Metz, France
[5] Holzinger Group, HCI-KDD, Institute for Medical Informatics/Statistics, Medical University Graz, Graz, Austria

Abstract. Convolutional Neural Networks (CNN) are very useful for fully automatic extraction of discriminative features from raw sensor data. This is an important problem in activity recognition, which is of enormous interest in ambient sensor environments due to its universality on various applications. Activity recognition in smart homes uses large amounts of time-series sensor data to infer daily living activities and to extract effective features from those activities, which is a challenging task. In this paper we demonstrate the use of the CNN and a comparison of results, which has been performed with Long Short Term Memory (LSTM), recurrent neural networks and other machine learning algorithms, including Naive Bayes, Hidden Markov Models, Hidden Semi-Markov Models and Conditional Random Fields. The experimental results on publicly available smart home datasets demonstrate that the performance of 1D-CNN is similar to LSTM and better than the other probabilistic models.

Keywords: Deep learning · Convolutional neural networks · 1D-CNN · LSTM · Activity recognition · Smart homes

1 Introduction

The advancement in sensing, networking and ambient intelligence technologies has resulted in emergence of smart environments and different services for a better quality of life and well being of the aging population. The aim are services providing comfort and security in their private space. Among them, the research in Smart Home (SH) has gained a lot of interest in the field of Ambient Assisted Living (AAL) technologies. The motivation behind the smart home research is the rapid increase in the world's aging population. According to the World Health Organization (WHO), the number of older people (aged 60 years or above) has

© Springer International Publishing AG 2017
A. Holzinger et al. (Eds.): Integrative Machine Learning, LNAI 10344, pp. 194–205, 2017.
https://doi.org/10.1007/978-3-319-69775-8_12

increased substantially in the past decade and expected to reach about 2 billion by 2050 [1].

The concept of smart homes gained popularity in early 2000s. Lutolf [2] defined smart home concept as the integration of different services within a home environment by using a common communication system. According to Satpathy [3] a smart home provides independence and comfort to the residents by using all mechanical and digital devices interconnected in a network and able to communicate with the user to create an interactive space.

Smart home equipped with simple, easy to install and low cost interconnected sensors are providing variety of services such as health care, well being, energy conservation by ensuring safety and security to the residents. Activity recognition in the home environment facilitates the remote monitoring for the purpose of detecting so called Activities of Daily Living (ADL), residents' behavior and their interaction with the smart environment. The large amount of data collected from the installed sensors is analyzed by employing machine learning models to detect meaningful features and abnormal behavioral patterns in ADLs. Several models have been proposed to recognize the activities inside smart homes using intrusive and non-intrusive approaches. Activity recognition by intrusive approaches is opposed to ethical aspects, e.g. devices such as video cameras, microphones in private environment raise privacy concern and therefore, unlikely to be accepted by the residents. On the other hand, non-intrusive approaches are preferable as they include simple and ubiquitous sensors to measure activities of the residents and the surroundings without hindering their privacy.

In the recent years, there has been extensive interest in deep learning in the field of image analysis [4], speech recognition [5] and sensor informatics [6]. Activity recognition using deep learning has several advantages in terms of system performance and flexibility. It provides an effective tool for extracting high-level feature hierarchies from high-dimensional sensory data which is useful for classification and regression tasks [7]. Deep learning models are based on learning representations from raw data and contain more than one hidden layer. The network learns many layers of non-linear information processing for feature extraction and transformation. Each successive layer uses the output from the previous layer as input. The well known deep learning models include Long Short Term Memory (LSTM) [8], Convolutional Neural Network (CNN) [9], Deep Belief Network (DBN) [10] and autoencoders [11].

In this work we exploit activity recognition using convolutional neural network model on publicly available smart homes dataset [12] which is an extension of our previous work of activity recognition using LSTM model [13]. The classification of the daily human activities such as cooking, bathing and sleeping is performed using temporal 1D-CNN model and evaluation of results has been carried out in comparison with LSTM and other machine learning algorithms such as Naive Bayes, Hidden Markov Model (HMM), Hidden Semi-Markov Model (HSMM) and Conditional Random Fields (CRF).

The paper is structured in different sections; the introduction is followed by Sect. 2 which presents an overview of existing work in activity recognition using various machine learning techniques in the field of AAL. Section 3 introduces Long Short-Term Memory and Convolutional Neural Network model. Section 4 describes the datasets that were used and explains the results. Finally, Sect. 5 discusses the outcomes of the experiments and suggestions for future work.

2   Related Work

The section is divided into three parts. The first part gives an overview of the existing smart home projects in the field of AAL. The second lists the available smart home datasets and the last part presents the existing work in activity recognition using machine learning techniques.

2.1   Existing Smart Homes

Several smart home projects have been implemented in the past decade, which use sensors for activity recognition inside the home environment. The Gator-tech smart house built by University of Florida contained smart appliances equipped with sensors such as smart blinds, smart refrigerator, smart stove which monitor user activities and provide services to the residents [14]. The Aware Home developed by Georgia Institute of Technology uses radio frequency identification (RFID) tags for the localization of the resident [15]. For the purpose of activity recognition, House_n project has been developed by Massachusetts Institute of Technology [16]. Various sensors have been installed to detect the routined activities such as toileting, bathing and grooming using supervised learning algorithms. The Center for Advanced Studies in Adaptive Systems (CASAS) introduced smart home in a box technology which is easy to install and provides various services with no customization and training [17]. Several other smart environment efforts have been demonstrated such as Easy Living project of Microsoft implements an intelligent environment to track and identify multiple residents through an active badge system [18]. In all of the smart home projects [19], activity recognition plays an important role.

2.2   Publicly Available Dataset

There has been several efforts to collect datasets from the sensors installed in the smart homes for human activity recognition. As these datasets are important for the research community since collecting real house annotated datasets is costly, time consuming and difficult to obtain.

The publicly available datasets are useful as they provide the baseline for the comparison of different machine learning algorithms. Eventually, it helps in collecting real house dataset using the baseline by identifying loopholes (if any) and corresponding improvement. The Table 1 summarizes the widely used publicly available smart home datasets.

Table 1. Publicly available smart home datasets

Dataset	Number of houses	Residents	Number of sensors	Number of activities
CASAS [17]	7	Multi	20 − 86	11
Kasteren [12]	3	Single	14 − 21	10 − 16
Ordonez [20]	2	Single	12	10 − 11
House_n [16]	2	Single	77 − 84	9 − 13
ARAS [21]	2	Multi	20	12 − 14
HIS [22]	1	Multi	20 − 30	7
OPPURTUNITY [23]	1	Multi	72	15 − 20

2.3   Activity Recognition in AAL

Activity recognition in smart home has been viewed as a promising approach to improve healthcare services and providing independent life to the older people. Activity recognition in SH has been broadly classified into two approaches: data driven and knowledge driven approaches. Data driven approaches use probabilistic and statistical models to learn from the datasets. Various supervised techniques have been widely used for classification of activity such as Naive Bayes [16], HMM [24], CRF and Support Vector Machine (SVM) [22]. The data driven approaches support modeling of uncertainty and temporal information but require large dataset to learn the model. Knowledge driven approaches do not require large datasets. Instead, these approaches use domain knowledge and prior heuristics to generate activity models, which make the model static and limited to specific representation of ADLs [25]; additionally, they cannot handle uncertainty and temporal parameters. Knowledge driven approaches for activity models and recognition are categorized into three types: Mining-based approach, Logic-based approach and Ontology-based approach.

The recent research has been focusing on activity recognition using unsupervised and semi-supervised machine learning algorithms to minimize the need of user annotation of activity datasets which requires a considerable amount of time and effort. The K-means clustering [26] is the most common as it performs best in temporal complexity and cluster set flexibility for large sensor datasets. The other clustering algorithms include hierarchical clustering, Density based clustering (DBSCAN) and Self-organizing map (SOM) which provides higher accuracy with random datasets [27]. However, the above clustering algorithms have some ambiguity in processing noise in the dataset. In addition, active learning techniques [28] have been used in the training process to improve accuracy in recognizing and forecasting activities in smart home.

3 Deep Learning

Nowadays, activity recognition using deep learning has become one of the most preferred techniques owing to its ability to learn data representations and classifiers. Their performance on different activity recognition tasks has been explored by researchers [29,30]. Deep architectures with multiple layers of Restricted Boltzmann Machines (RBM) handle binary sensory data and use DBN-ANN and DBN-R algorithms for human behavior prediction [31]. Convolutional neural networks [32] are type of deep neural network (DNN) which use convolutions over the input layer to compute the output. Each layer applies different filters to extract hierarchical features, dependency, translation equi-variance of data and automates feature learning, which make it suitable to use for time series raw sensor data. The CNN model has performed well in extracting features and recognizing activities from the raw sensor data [33] and video frames in comparison to the other machine learning approaches on publicly available datasets [34]. In this paper, we performed 1D-CNN on publicly available smart home datasets.

3.1 LSTM Model

LSTM, proposed by [35], is a recurrent neural network architecture which is capable of learning long term dependencies. LSTM has been developed in order to deal with gradient decay or gradient blow-up problems and can be seen as a deep neural network architecture when unrolled in time. The LSTM layer's main component unit is called memory cell. A memory cell is composed of four main elements: an input gate, a neuron with self-recurrent connection, a forget gate and an output gate [13]. The input provided to the LSTM controls the operations to be performed by the gates in the memory cell: write (input gate), read (output gate) and reset (forget gate). Following equations explain the way a layer of memory cells is updated at each timestep t.

$$i_t = \sigma(W_{xi}x_t + W_{hi}h_{t-1} + W_{ci}c_{t-1} + b_i),$$
$$f_t = \sigma(W_{xf}x_t + W_{hf}h_{t-1} + W_{cf}c_{t-1} + b_f),$$
$$o_t = \sigma(W_{xo}x_t + W_{ho}h_{t-1} + W_{co}c_t + b_o),$$
$$c_t = f_tc_{t-1} + i_t \tanh(W_{xc}x_t + W_{hc}h_{t-1} + b_c),$$
$$h_t = o_t \tanh c_t,$$

where W_i, W_f, W_o are the weight matrix and x_t is the input to the memory cell layer at time t, σ being the sigmoid and tanh is the hyperbolic tangent activation function. The terms i, f and o are the input gate, forget gate and output gate. The term c represents the memory cell and b_i, b_f, b_c and b_o are bias vectors.

Figure 1 illustrates an LSTM single cell layer at time t where x_t,h_t and y_t are the input, hidden and output state.

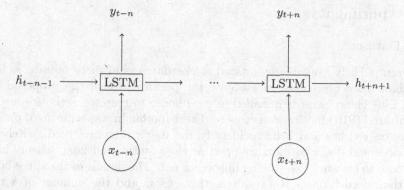

Fig. 1. Illustrations of an LSTM network with x being the binary vector for sensor input and y being the activity label prediction of the LSTM network.

3.2   CNN Model

Convolutional neural network is a type of deep neural network, consists of multiple hidden layers which can be either convolutional, pooling or fully connected. A single convolutional layer of CNN extracts features from the input signal through convolution operation of the signal with a kernel. The activation of a unit in a CNN represents the output of the convolution of the kernel with the input signal. CNNs are able to learn hierarchical data representations for fast feature extraction and classification. The CNN model has advantages when used for activity recognition task [36]. It can capture local dependencies of the activity signal and preserves feature scale invariant, thus able to capture variations in the similar activity efficiently through feature extraction. Figure 2 shows the structure of CNN for Activity Recognition.

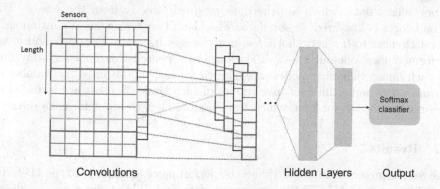

Fig. 2. CNN for activity recognition

1D temporal convolutional model used in this work, has four layers: (1) an input layer, (2) convolution layer with multiple feature widths and feature map, (3) fully connected layer and (4) the output layer.

4   Experiments

4.1   Dataset

Kasteren publicly available annotated sensor datasets of three houses, as listed in Table 1 have been used to evaluate the performance of the proposed approach. The binary sensors installed in each house to monitor activities are passive infrared (PIR) motion detectors to detect motion in a specific area, pressure sensors on couches and beds to identify the user's presence, reed switches on cupboards and doors to measure open or close status and float sensors in the bathroom to measure toilet being flushed or not. The details of the three houses with the information of the resident, the sensors and the number of activity labels are provided in Table 2.

Table 2. Details of the datasets.

	House A	House B	House C
Age	26	28	57
Gender	Male	Male	Male
Setting	Apartment	Apartment	House
Rooms	3	2	6
Duration	25 days	14 days	19 days
Sensors	14	23	21
Activities	10	13	16
Annotation	Bluetooth	Diary	Bluetooth

The data in the experiments are represented in two different forms. The first is raw sensor data, which are the data received directly from the sensor. The second form is last-fired sensor data. The last firing sensor gives continuously 1 and changes to 0 when another sensor changes its state. For each house, we performed leave-one-out cross validation and repeated this for every day and for each house. Separate models are trained for each house since the number of sensors varies and a different user resides in each house. Sensors are recorded at one-minute interval for 24 h, which totals in 1440 length input for each day.

4.2   Results

The results presented in Table 3 show the performance (accuracy) of the 1D-CNN model together with LSTM on raw sensor data and Table 4 shows the results of the last-fired sensor data in comparison with the results of Naive Bayes, HMM, HSMM and CRF [12]. We calculated accuracy of the model, which represents the correctly classified activities in each time. For the LSTM model, a time slice of (70) with hidden state size (300) are used. We implemented 1D(temporal) convolution with a time slice of (15). 128 filters are used for each layer and 1D kernel sizes were 5, 5, 3, 3, 3, 3 with a fully connected layer of 128 in the end. Dropout of 0.5 is used in order to reduce the overfitting in the data. We also tested longer timeslices but they tend to overfit considerably. Adam method [37] is used with a learning rate of 0.0004 for optimization of the networks and Tensorflow library of Python has been used to implement the CNN and LSTM network. The training took place on a Titan X GPU and the time required to train one day for one house is 4 min for CNN and approximately 30 min for LSTM, but training time differs amongst the houses. Since different houses have different number of days of data, we calculated the average accuracy amongst all days. The training is performed using a single GPU but the trained models can be used for inference without losing performance when there is no GPU.

Table 3. Results of raw sensor data

Model	House A	House B	House C
1D-CNN*	88.2 ± 8.6	79.4 ± 20.1	49.2 ± 25.6
LSTM**	**89.8 ± 8.2**	**85.7 ± 14.3**	**64.22 ± 21.9**
Naive Bayes	77.1 ± 20.8	80.4 ± 18.0	46.5 ± 22.6
HMM	59.1 ± 28.7	63.2 ± 24.7	26.5 ± 22.7
HSMM	59.5 ± 29.0	63.8 ± 24.2	31.2 ± 24.6
CRF	89.8 ± 8.5	78.0 ± 25.9	46.3 ± 25.5

*Current work, **Previous work [13]

Each model for each house is trained with leave-one-day out strategy. If house has k days of data k-1 days are used to train and 1 day is used to test and this processed is repeated for each day. In order to compare models average accuracy with variance are calculated. Table 3 shows the average accuracy with the variance of accuracies of different models on raw data from three different houses. Among all the models, the LSTM performs the best for all three datasets and 1D-CNN performs second best. In House B and House C, LSTM improves the best result significantly especially on House C where the improvement is approximately 40% from CRF and 30% from CNN.

Table 4. Results of last-fired sensor data

Model	House A	House B	House C
1D-CNN*	95.3 ± 2.8	86.8 ± 12.7	86.23 ± 12.4
LSTM**	95.3 ± 2.0	88.5 ± 12.6	85.9 ± 10.6
Naive Bayes	95.3 ± 2.8	86.2 ± 13.8	87.0 ± 12.2
HMM	89.5 ± 8.4	48.4 ± 26.0	83.9 ± 13.9
HSMM	91.0 ± 7.2	67.1 ± 24.8	84.5 ± 13.2
CRF	**96.4 ± 2.4**	**89.2 ± 13.9**	**89.7 ± 8.4**

*Current work, **Previous work [13]

Table 4 shows the accuracy on last fired data from three different houses. The 1D-CNN matches the best performance achieved by CRF in case of House A but drops slightly in case of House B and C. In comparison to LSTM, 1D-CNN performs similar except a slight decrease in case of House B. It is also important to notice the high variance in all models. Variance is halved for the last-fired sensor data compared to raw sensor data.

5   Conclusions and Future Work

In this work, we used deep learning techniques for activity recognition from raw sensory inputs in smart home environment. As data preprocessing and feature engineering is expensive for real world applications especially in AAL environment, the prediction from raw input data can eliminate most of the feature engineering efforts performed by humans. Deep learning models (1D-CNN and LSTM) lead to significant improvement in performance, especially on raw data in comparison to existing probabilistic models such as Naive Bayes, HMM, HSMM and CRF. In case of last fired data, both deep learning models (1D-CNN and LSTM) match the best performance of existing models. Although, LSTM gives better performance than CNN in general cases, but when it comes to training times CNN is much faster than LSTM based models. The selection of CNN over LSTM can help in reducing time to design a prototype and see if there is a temporal dependence between input and output. In addition, CNN also helps in evaluating performance of different architectures to achieve best results which could be very time consuming with LSTM.

In general, there are many future research directions of deep learning approaches in medical applications and specifically, in ambient assisted living scenarios. One problem in the medical domain is that the deep learning approaches are so-called black-box approaches, thus are lacking transparency. However, in the medical domain trust and acceptance among end-users is of eminent importance. Consequently, a big research challenge will emerge through rising legal and privacy aspects, e.g. with the new European General Data Protection Regulations [38], it will become a necessity to explain why a decision has been made [39]. An interesting emerging field in the ever-increasing complexity and large number of heterogeneous sensors (in sensor networks of ambient assistant living) is to combine deep learning approaches with graphical and topological approaches [40], which leads to geometric deep learning [41]. This is just to outline the enormous potential of future research in the area of deep learning applied to ambient assisted living - as part of health systems.

Nevertheless, the immediate future work will be focusing on combining CNN with LSTM to utilize the fast training time with high accuracy. More detailed architectures search for only CNN based models is also planned to be performed. In order to avoid overfitting, different methods will be investigated for a better generalization. It could also be interesting to investigate the high variance experienced in some cases.

Acknowledgement. This work has been funded by the European Union Horizon2020 MSCA ITN ACROSSING project (GA no. 616757). The authors would like to thank the members of the project's consortium for their valuable inputs.

References

1. United Nations, Department of Economic and Social Affairs: United nations department of economic and social affairs, population division: World population ageing 2015 (2015)
2. Lutolf, R.: Smart home concept and the integration of energy meters into a home based system. In: Seventh International Conference on Metering Apparatus and Tariffs for Electricity Supply, 1992, IET, pp. 277–278 (1992)
3. Satpathy, L.: Smart housing: technology to aid aging in place: new opportunities and challenges. Ph.D. thesis, Mississippi State University (2006)
4. He, K., Zhang, X., Ren, S., Sun, J.: Deep residual learning for image recognition. In: Proceedings of the IEEE Conference on Computer Vision and Pattern Recognition, pp. 770–778 (2016)
5. Deng, L., Li, J., Huang, J.T., Yao, K., Yu, D., Seide, F., Seltzer, M., Zweig, G., He, X., Williams, J., et al.: Recent advances in deep learning for speech research at microsoft. In: 2013 IEEE International Conference on Acoustics, Speech and Signal Processing (ICASSP), pp. 8604–8608. IEEE (2013)
6. Längkvist, M., Karlsson, L., Loutfi, A.: A review of unsupervised feature learning and deep learning for time-series modeling. Pattern Recogn. Lett. **42**, 11–24 (2014)
7. Salakhutdinov, R.: Learning deep generative models. Ann. Rev. Stat. Appl. **2**, 361–385 (2015)
8. Sutskever, I., Vinyals, O., Le, Q.V.: Sequence to sequence learning with neural networks. In: Advances in Neural Information Processing Systems, pp. 3104–3112 (2014)
9. Matsugu, M., Mori, K., Mitari, Y., Kaneda, Y.: Subject independent facial expression recognition with robust face detection using a convolutional neural network. Neural Netw. **16**(5), 555–559 (2003)
10. Hinton, G.E., Osindero, S., Teh, Y.W.: A fast learning algorithm for deep belief nets. Neural Comput. **18**(7), 1527–1554 (2006)
11. Hinton, G.E., Salakhutdinov, R.R.: Reducing the dimensionality of data with neural networks. Science **313**(5786), 504–507 (2006)
12. Kasteren, T.L., Englebienne, G., Kröse, B.J.: Human activity recognition from wireless sensor network data: benchmark and software. In: Chen, L., Nugent, C., Biswas, J., Hoey, J. (eds.) Activity Recognition in Pervasive Intelligent Environments, pp. 165–186. Atlantis Press, Amsterdam (2011)
13. Singh, D., Merdivan, E., Psychoula, I., Kropf, J., Hanke, S., Geist, M., Holzinger, A.: Human Activity recognition using recurrent neural networks. In: Holzinger, A., Kieseberg, P., Tjoa, A.M., Weippl, E. (eds.) CD-MAKE 2017. LNCS, vol. 10410, pp. 267–274. Springer, Cham (2017). doi:10.1007/978-3-319-66808-6_18
14. Helal, S., Mann, W., El-Zabadani, H., King, J., Kaddoura, Y., Jansen, E.: The gator tech smart house: a programmable pervasive space. Computer **38**(3), 50–60 (2005)
15. Kidd, C.D., et al.: The aware home: a living laboratory for ubiquitous computing research. In: Streitz, N.A., Siegel, J., Hartkopf, V., Konomi, S. (eds.) CoBuild 1999. LNCS, vol. 1670, pp. 191–198. Springer, Heidelberg (1999). doi:10.1007/10705432_17

16. Tapia, E.M., Intille, S.S., Larson, K.: Activity recognition in the home using simple and ubiquitous sensors. In: Ferscha, A., Mattern, F. (eds.) Pervasive 2004. LNCS, vol. 3001, pp. 158–175. Springer, Heidelberg (2004). doi:10.1007/978-3-540-24646-6_10

17. Cook, D.J., Crandall, A.S., Thomas, B.L., Krishnan, N.C.: Casas: a smart home in a box. Computer **46**(7), 62–69 (2013)

18. Brumitt, B., Meyers, B., Krumm, J., Kern, A., Shafer, S.: EasyLiving: technologies for intelligent environments. In: Thomas, P., Gellersen, H.-W. (eds.) HUC 2000. LNCS, vol. 1927, pp. 12–29. Springer, Heidelberg (2000). doi:10.1007/3-540-39959-3_2

19. Alam, M.R., Reaz, M.B.I., Ali, M.A.M.: A review of smart homespast, present, and future. IEEE Trans. Syst. Man Cybern. Part C (Applications and Reviews) **42**(6), 1190–1203 (2012)

20. Ordóñez, F.J., de Toledo, P., Sanchis, A.: Activity recognition using hybrid generative/discriminative models on home environments using binary sensors. Sensors **13**(5), 5460–5477 (2013)

21. Alemdar, H., Ertan, H., Incel, O.D., Ersoy, C.: Aras human activity datasets in multiple homes with multiple residents. In: Proceedings of the 7th International Conference on Pervasive Computing Technologies for Healthcare, ICST (Institute for Computer Sciences, Social-Informatics and Telecommunications Engineering), pp. 232–235 (2013)

22. Fleury, A., Noury, N., Vacher, M.: Supervised classification of activities of daily living in health smart homes using svm. In: Annual International Conference of the IEEE Engineering in Medicine and Biology Society, EMBC 2009, pp. 6099–6102. IEEE (2009)

23. Roggen, D., Calatroni, A., Rossi, M., Holleczek, T., Förster, K., Tröster, G., Lukowicz, P., Bannach, D., Pirkl, G., Ferscha, A., et al.: Collecting complex activity datasets in highly rich networked sensor environments. In: 2010 Seventh International Conference on Networked Sensing Systems (INSS), pp. 233–240. IEEE (2010)

24. Monekosso, D.N., Remagnino, P.: Anomalous behavior detection: supporting independent living. In: Monekosso, D., Remagnino, P., Kuno, Y. (eds.) Intelligent Environments, pp. 33–48. Springer, London (2009)

25. Chen, L., Hoey, J., Nugent, C.D., Cook, D.J., Yu, Z.: Sensor-based activity recognition. IEEE Trans. Syst. Man Cybern. Part C (Applications and Reviews) **42**(6), 790–808 (2012)

26. Lapalu, J., Bouchard, K., Bouzouane, A., Bouchard, B., Giroux, S.: Unsupervised mining of activities for smart home prediction. Procedia Comput. Sci. **19**, 503–510 (2013)

27. Li, C., Biswas, G.: Unsupervised learning with mixed numeric and nominal data. IEEE Trans. Knowl. Data Eng. **14**(4), 673–690 (2002)

28. Longstaff, B., Reddy, S., Estrin, D.: Improving activity classification for health applications on mobile devices using active and semi-supervised learning. In: 2010 4th International Conference on- Pervasive Computing Technologies for Healthcare (PervasiveHealth), pp. 1–7. IEEE (2010)

29. Hammerla, N.Y., Halloran, S., Ploetz, T.: Deep, convolutional, and recurrent models for human activity recognition using wearables. arXiv preprint arXiv:1604.08880 (2016)

30. Yang, J., Nguyen, M.N., San, P.P., Li, X., Krishnaswamy, S.: Deep convolutional neural networks on multichannel time series for human activity recognition. IJCAI, 3995–4001 (2015)

31. Choi, S., Kim, E., Oh, S.: Human behavior prediction for smart homes using deep learning. In: RO-MAN, 2013 IEEE, pp. 173–179. IEEE (2013)
32. Goodfellow, I., Bengio, Y., Courville, A.: Deep Learning. MIT Press (2016). http://www.deeplearningbook.org
33. Chen, Y., Xue, Y.: A deep learning approach to human activity recognition based on single accelerometer. In: 2015 IEEE International Conference on Systems, Man, and Cybernetics (SMC), pp. 1488–1492. IEEE (2015)
34. Geng, C., Song, J.: Human action recognition based on convolutional neural networks with a convolutional auto-encoder. In: 5th International Conference on Computer Sciences and Automation Engineering (ICCSAE 2015) (2015)
35. Hochreiter, S., Schmidhuber, J.: Long short-term memory. Neural Comput. **9**(8), 1735–1780 (1997)
36. Zeng, M., Nguyen, L.T., Yu, B., Mengshoel, O.J., Zhu, J., Wu, P., Zhang, J.: Convolutional neural networks for human activity recognition using mobile sensors. In: 2014 6th International Conference on Mobile Computing, Applications and Services (MobiCASE), pp. 197–205. IEEE (2014)
37. Kingma, D., Ba, J.: Adam: a method for stochastic optimization. arXiv preprint arXiv:1412.6980 (2014)
38. Barnard-Wills, D.: The technology foresight activities of european union data protection authorities. Technol. Forecast. Soc. Change **116**, 142–150 (2017)
39. Holzinger, A., Plass, M., Holzinger, K., Crisan, G.C., Pintea, C.M., Palade, V.: A glass-box interactive machine learning approach for solving np-hard problems with the human-in-the-loop. arXiv:1708.01104 (2017)
40. Holzinger, A.: On topological data mining. In: Holzinger, A., Jurisica, I. (eds.) Interactive Knowledge Discovery and Data Mining in Biomedical Informatics. LNCS, vol. 8401, pp. 331–356. Springer, Heidelberg (2014). doi:10.1007/978-3-662-43968-5_19
41. Bronstein, M.M., Bruna, J., LeCun, Y., Szlam, A., Vandergheynst, P.: Geometric deep learning: going beyond euclidean data. IEEE Sig. Process. Mag. **34**(4), 18–42 (2017)

Author Index

Printed in the United States
By Bookmasters